# Customize Your Phone

## 15 Electronic Projects

No. 3054
$19.95

# Customize Your Phone

## 15 Electronic Projects

Steve Sokolowski

TAB BOOKS Inc.
Blue Ridge Summit, PA

FIRST EDITION
FIRST PRINTING

Copyright © 1989 by TAB BOOKS Inc.
Printed in the United States of America

Reproduction or publication of the content in any manner, without express
permission of the publisher, is prohibited. No liability is assumed with respect to
the use of the information herein.

Library of Congress Cataloging in Publication Data

Sokolowski, Steve.
    Customize your phone.

    Includes index.
    1.   Telephone—Equipment and supplies—Design and
construction—Amateurs' manuals.   I.   Title.
TK9951.S65   1988                 621.386                 88-24795
ISBN 0-8306-9054-9
ISBN 0-8306-9354-8 (pbk.)

TAB BOOKS Inc. offers software for sale. For information and
a catalog, please contact TAB Software Department, Blue Ridge
Summit, PA 17294-0850.

Questions regarding the content of this book
should be addressed to:

    Reader Inquiry Branch
    TAB BOOKS Inc.
    Blue Ridge Summit, PA 17294-0214

# Contents

## PROJECT LIST

| | NAME |
|---|---|
| 1 | Tele-Guard |
| 2 | Telephone Hold Button |
| 3 | Telephone Melody Ringer |
| 4 | Telephone Tone Ringer |
| 5 | Automatic Telephone Recorder |
| 6 | Telephone Call Indicator |
| 7 | Ring Detector |
| 8 | Conference Caller |
| 9 | Telephone Lock |
| 10 | Telephone Intercom |
| 11 | Telephone Line Tester |
| 12 | Telephone Amplifier |
| 13 | Speaker Phone |
| 14 | Appliance Controller |
| 15 | Animated Telephone Ringer |

# Acknowledgments

At this time, I would like to thank the people of Corinth Telecommunications Corporation located in Corinth, Mississippi, for their help in the preparation of this book, and for allowing me to reprint a number of illustrations that appeared in their Telephone Apparatus Practice Manual, 1987 edition.

I would also like to thank the people and engineers of Del-Phone Industries Inc. for their help in designing many of the telephone projects that appear within.

Without the help of all these people, this book would still be an idea, without any chance of being a reality.

# Introduction

A court decision killed the telephone monopoly that AT&T enjoyed for a number of years. The court not only broke up AT&T, but also gave consumers the right to purchase and connect telephone equipment to our once off-limit telephone lines. Manufacturers saw an opportunity arise and they did something to fill the void AT&T left.

You can now go to any large retail, electronic, or telephone specialty store to purchase all sorts of telephone equipment. Today, you can do something that only a short time ago made the engineers of AT&T see double; that is, connect outside purchased equipment to *their* telephone lines yourself.

With this book, you can take advantage of this courtroom decision. You can build a large variety of telephone-related equipment. The projects in this book not only give you experience in building electronic devices, but provide a means of enhancing the use of your telephone.

You will find projects that range from a telephone bug detector (Tele-Guard) to a musical hold button, to a project that allows you to control any household appliance by using a standard tone telephone which can be located anywhere in the world. All project designs, as well as instructional text, are presented in easy to understand language.

Even if you have never constructed electronic projects, you will be taken step-by-step through a chapter, explaining how to select the right components for a circuit, fabrication of a printed circuit board, and the do's and don'ts of soldering. When needed, total circuit theory as well as printed circuit board artwork is explained in clear and easy to understand terms.

You can count on a thorough introduction to telephone electronics. You will be taken on a journey that examines the internal operation, as well as how telephones transform a speech pattern (your voice) to electrical signals, and how these signals are reconstructed to a recognizable voice.

Detailed instructional theory on the operation and internal circuitry of a rotary and tone dial is also discussed. Circuit diagrams of standard tone dials are presented courtesy of the Corinth Telecommunication Corporation. Learn how and why the older LC type tone dials are being slowly removed from service to make way for the newer frequency synthesized tone dials.

Be prepared to journey through the fascinating world of telephone communications: the hows, whys, dos, and don'ts of connecting homemade circuits to a telephone line.

Thumb through this book and discover how easy it is to develop, construct, and install telephone equipment—equipment that will not only give you pride in having built a working electronic project, but also the sense of accomplishment that you did it yourself (with a little help from this book).

# 1   What Is a Telephone?

THE TELEPHONE HAS PROBABLY BEEN THE MOST WIDELY USED ELECTRONIC IN-strument since its arrival over a century ago. In the United States, there is an estimated 100 million telephones now in use.

It staggers the mind to think that any two of these telephones can be connected together to allow the human voice or computer data to pass between them using a pair of copper wires or a beam of light (fiberoptic technology). Yet this huge network access can be accomplished by an unskilled operator: you. Just by picking up the telephone receiver the sequence of complicated network events comes alive. It was all made possible in 1876 by Alexander Graham Bell when he spilled acid on his trousers then yelled-out the now famous quote "Mr. Watson, come here, I want you."

Over the past few years, the telephone, like the one found in your home, has come through monumental changes. Telephones are now highly sophisticated electronic devices using a wide variety of integrated circuits as well as computer chips.

The shape and style of the telephone might have changed, but the basic functions have not. Here is a list of the most important functions:

☐ When you lift the handset, the telephone signals to the local exchange that a call is to be made.

☐ It indicates to you that the local exchange is ready to process the call by allowing you to receive a tone. This tone is called the dial tone.

☐ It sends the number of the telephone you wish to be connected to by pressing numbered buttons or rotating a plastic dial.

☐ It indicates the status of the call in process by receiving a sequence of tones (ringing, busy tone, etc.)

☐ It indicates that a person is trying to get in touch with you by the ringing of a bell or other audible signal.

☐ It changes a speech pattern of a calling party to electrical signals easily passed through copper wires, and converts it back into an intelligible audio pattern at the receiving instrument.

☐ It automatically adjusts its internal operation to reflect changes in voltage supplied to it.

☐ It signals to the local exchange that the call is complete when the handset is placed back on the telephone cradle.

Fig. 1-1. A standard 500-type telephone set. (Courtesy of Corinth Telecommunications Corp.)

One of the more commonly used telephones today is the 500 type unit. This is a basic desk model with a rotary dial. This dial produces pulses that correspond to the number of the telephone you wish the network to connect you to.

The counterpart to the 500 type telephone is the 2500. The 2500 telephone, like the 500, is a standard desk telephone, but uses a technology that was first introduced to the public at the New York Worlds Fair back in the '60s.

Instead of pulsing-out the desired telephone number, the 2500 converts this number to a series of specially selected tones, which the local network converts back into the desired dialed number.

Not all telephones manufactured are desk types. Telephones that are mounted on a wall are also produced. They are called, as the name implies, wall phones.

A wall-mounted counterpart to the 500-type rotary desk telephone is the 554 model. For the tone dial enthusiast, the 2500 desk telephone has two substitutes: the 3554 and the more slender, compact, 2554 model.

Fig. 1-2. A standard 2500-type telephone set. (Courtesy of Corinth Telecommunications Corp.)

Fig. 1-3. A standard 554-type wall telephone. **(Courtesy of Corinth Telecommunications Corp.)**

Whichever model telephone you own, the operating functions still remain the same. Refer to Figs. 1-1, 1-2, 1-3, 1-4, and 1-5 for illustrations of the most common home telephones.

## TELEPHONE BASICS

Figure 1-6 is a schematic illustrating how sound is transformed into electrical signals that travel through a pair of wires and are converted back at the receiving end.

To understand how a telephone operates, you must understand what sound is. Sound in reality is a variation in air pressure. It is this air pressure that vibrates

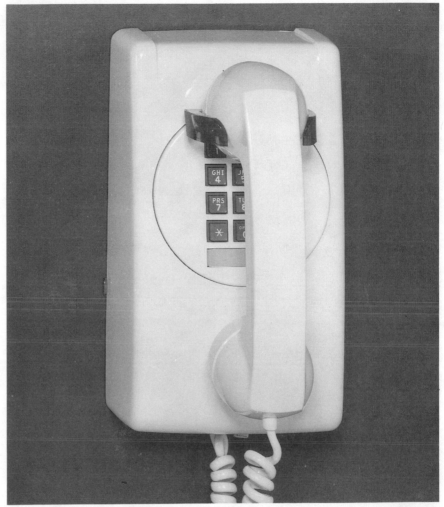

Fig. 1-4. A standard 3554-type tone dial telephone. **(Courtesy of Corinth Telecommunications Corp.)**

the diaphragm of a telephone transmitter. The diaphragm in turn applies this pressure to expand and condense carbon granules located within the transmitter. By expanding and condensing these granules, the resistance of the transmitter changes in proportion to the applied air pressure generated by speaking into the telephone handset. When a battery (called the talk battery) is connected to the transmitter, the varying resistance of the transmitter allows a varying current to flow. An exact reproduction of the sound created by your voice is converted into an electrical signal.

Sound is reproduced in a telephone by the receiver section of the handset (uppermost section of the handset). The receiver is an electromagnet with a metal diaphragm attached to it. Variations in the electrical current within the

Fig. 1-5. A standard 2554-type wall tone dial telephone. **(Courtesy of Corinth Telecommunications Corp.)**

coil of wire that makes up the electromagnet attracts and repels the metal diaphragm in proportion to the applied varying current. When this diaphragm vibrates, varying air pressure is created. This air pressure is then interpreted by your ear as sound.

You can easily understand this concept by bringing a telephone handset to your ear and blowing into the transmitter. You will be able to hear this sound in the receiver section. The sound of your own voice in the handset is called *side tone*. By hearing your own voice in the receiver, it prevents you from using an abnormally loud voice when speaking on the telephone. Figure 1-7 is an exploded view of a typical handset.

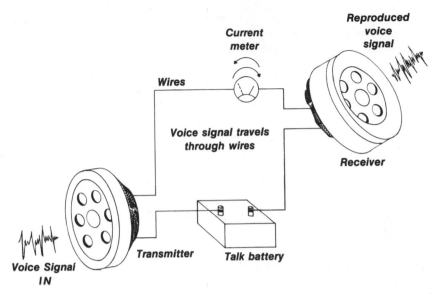

Fig. 1-6. An Audio Signal is converted into electrical pulses then back again.

If only two telephones are connected, and they are within a short distance from each other, a telephone can consist of just two handsets connected to a talk battery. This is far from reality, however. Say you made a call to a friend who lives across the street from you. Your voice signal will be very loud to your friend. He or she might have to hold the receiver away from their ear. Now if you make a call using the same telephone to another friend 100 miles away, the volume of your voice will be very low.

To eliminate this signal variation between a local and a long distance call, designers included a component called a *varistor*. This component is a voltage variable resistor. If the voltage across this varistor is high, the resistance of this component will be low. In reverse, if the voltage is low, the internal resistance of the varistor will be high.

Now if you make the same call to a friend across the street, the varistor will reduce the amount of signal produced by the transmitter, but if you make a long distance call, the volume of your voice on the second call will be the same as the volume received by your neighbor across the street.

Figure 1-8 is a schematic of a basic 500 type of telephone. RV1, a varistor, suppresses dialing clicks that are generated by the rotary dial. This varistor is soldered directly to the screw terminals of the telephone receiver-element found in the handset.

The balancing network, comprised of RV2, C2, C3, R2, and the induction coil, provides simultaneous two-way conversation using a two-wire system. Capacitor C1 and resistor R1 make up a dial pulse-filter, which is used to suppress

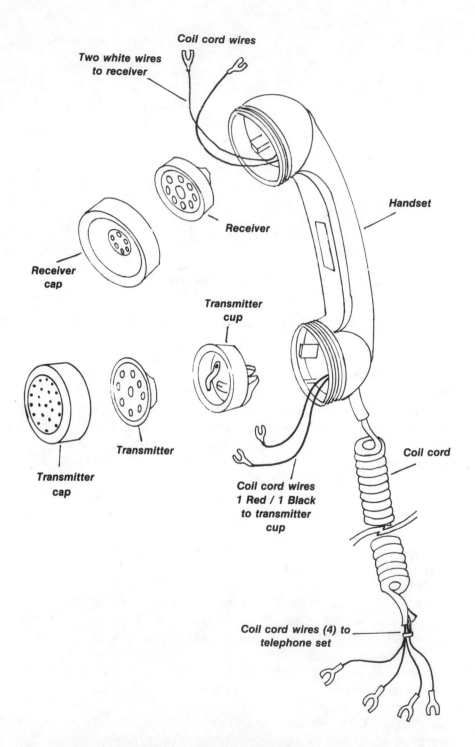

Coil cord wires

Two white wires
to receiver

Handset

Receiver

Receiver
cap

Transmitter
cup

Transmitter

Transmitter
cap

Coil cord

Coil cord wires
1 Red / 1 Black
to transmitter
cup

Coil cord wires (4) to
telephone set

Fig. 1-7. Exploded view of a telephone handset.

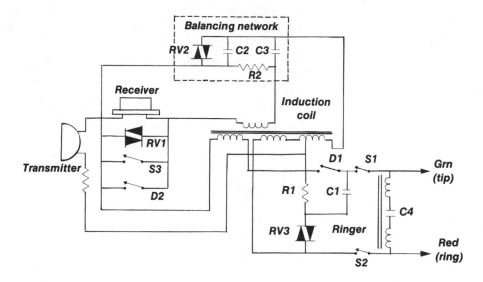

Fig. 1-8. A schematic of a standard telephone.

radio interference when dialing a telephone number. Varistors RV2 and RV3 are used to reduce the efficiency of the transmitter on local calls. These varistors maintain satisfactory transmission volume on both local, and long distance calls.

Most home-bound telephones have four wires coming out of them. These wires are colored green (Tip), red (Ring), yellow, and black, and are contained within a cable called a line cord. This line cord is then connected to a junction box (or connection block) located at the baseboard in a room. This junction box is a termination point, where your telephone is connected to the incoming telephone line through a lightning arrestor. This telephone line runs to your house from a nearby telephone pole. Figure 1-9 represents a typical telephone installation.

For a telephone to operate correctly, usually all one must do is connect the green and red wires of the telephone to the associated green and red wires within the connection block. Other types of installations require you to connect the yellow and green wires of the telephone line cord to the green wire in the connection block. This depends on the physical wiring inside the telephone. Some manufacturers connect these two wires inside the telephone; others require this connection be made outside the telephone at the connection block.

## HOW A TELEPHONE RINGS

For a telephone to ring, a voltage between 60 and 90 volts with a frequency of about 20 Hz is applied to the internal bell via the green and red (tip/ring) wires. This bell (or ringer) consists of two coils of wire with a capacitor in series (C4 in Fig. 1-8). When the high ringing voltage is present, a magnetic field

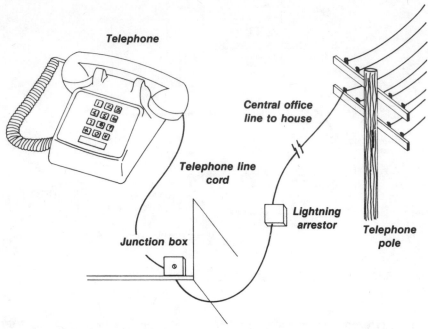

**Telephone**

**Central office
line to house**

**Telephone line
cord**

**Lightning
arrestor**

**Telephone
pole**

**Junction box**

Fig. 1-9. How a telephone is connected to the central office lines.

is generated, attracting and repelling a metal arm. The arm is allowed to strike two gongs: one at its left, and the second to its right. This clapping produces the ringing sound that is heard when a call is received. On large style telephones (500, 2500, 554 etc.), a two gong type ringer is used (138BA ringer). See Fig. 1-10. This bell produces a very loud ringing sound that can be controlled somewhat by moving one of the gongs closer to the vibrating arm. On a 3554 and 554 wall telephone, this adjustment is in a form of a lever located at the bottom of the telephone that can be pushed from side to side. On a 500 or 2500 desk telephone, a single adjustment is located on the underside of the instrument.

For smaller telephones, like the 2554, a bell with one gong is used (148 BA ringer). See Fig. 1-11. The volume of this type of ringer is considerably lower than the 138 BA ringer, but loudness is sacrificed for its compact size.

Both the 138 and the 148BA ringers are industrial standards and can be placed into operation anywhere in the country using a ring frequency of 20 Hz, which these two bells were designed to do. Figure 1-12 is an electrical diagram of the 138 and 148BA ringers. Note that the 148 ringer has one coil while the 138 has a two-coil ringing system.

Figure 1-8 illustrates what happens when the handset is lifted and a telephone can access the sophisticated switching equipment located at your local telephone exchange. When the handset is removed from the cradle, switches S1 and S2 close, while S3 opens. These three switches are contained in an assembly known

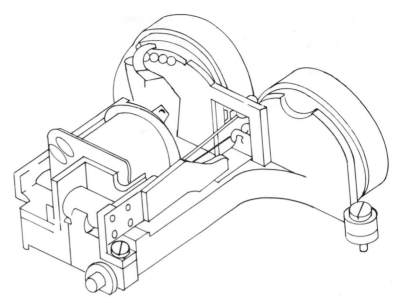

Fig. 1-10. A typical 2 gong telephone ringer (130BA ringer).

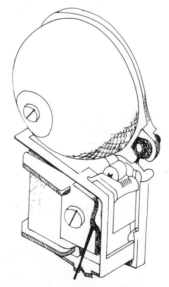

Fig. 1-11. A typical single gong telephone ringer (148BA ringer).

as a *hookswitch* (or switch-hook). Upon closing S1 and S2, the telephone is placed across the incoming line. The 48 volts (talk battery), which is present across the line (tip/ring) while on hook, now drops to about 5 volts. At this time a tone is generated by the local exchange indicating that the central office (where all telephone switching equipment is located) is ready to accept dialing and subsequent connection to the desired location. This tone is called the *dial tone*.

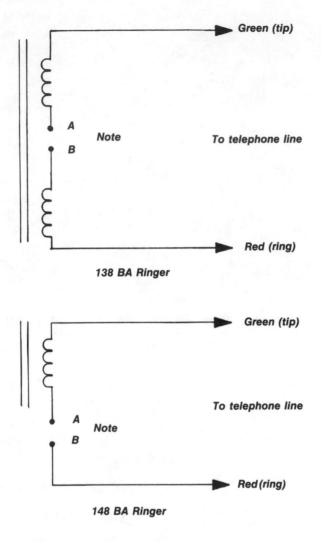

138 BA Ringer

148 BA Ringer

*Note: For proper operation, connect points A   and B to a .47 micro farad capacitor (200 volts)*

Fig. 1-12. An electronic schematic of both a 130 and 148BA ringers.

## ROTARY DIAL

To indicate to the central office the desired telephone number, a means of generating a *pulse-train* (continuous number of on and off connections made within a specific amount of time) is needed. This invention was first introduced in 1895 and is still in use today. It is called the *rotary dial*. (See Fig. 1-13.) The dial is a device that the user rotates with their finger to a predetermined position. When the finger is removed, the dial returns to its resting position.

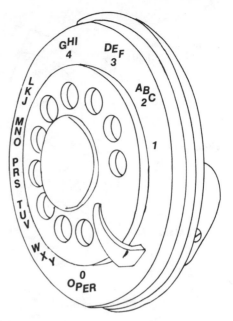

Fig. 1-13. A typical rotary dial.

While returning, the dial opens and closes a small switch (D1 in Fig. 1-14) at a rate of 10 times per second. The number of switch openings and closings corresponds to the number (or letter) the dial was turned to before being released? For an example, say you indicate, by using the Rotary Dial, the desired number is #5, the switch, D1, will open and close five times. If you indicated the number 9 on the dial, D1 would open and close nine times. Equipment at the central office counts each pulse then connects your telephone to the correct number dialed.

To prevent loud clicking in the telephone receiver as switch D1 opens and closes, switch D2 (wired across the telephone receiver) closes, shorting the receiver during the dialing interval. Refer to Fig. 1-14 to see how switches D1 and D2 are interconnected using the rotary dial.

## PUSH-BUTTON DIALING

Today, more and more telephone sets are manufactured using the newer method of audio tones to send the telephone number to the central office. The 2500 model telephone uses this type of dialing. Instead of a rotary dial, the tone phones have a push-button keypad with 12 keys for the numbers 0 to 9 and the symbols * (asterisk) and # (octothorpe). Figure 1-15 is an illustration of a common tone dial, the 32opg type. Pressing one of the keys causes an electronic circuit in the keypad to generate two output tones that represent the number.

To represent telephone numbers, eight frequencies in the 700 to 1700 Hz range comprise a four-by-three code designed for push-button dialing. There are four low-band and three high-band tones, as illustrated in Fig. 1-16.

14

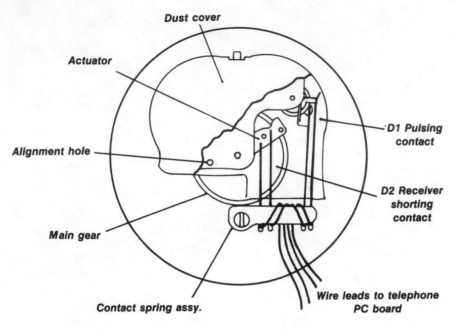

Fig. 1-14. Mechanical switches of a rotary dial.

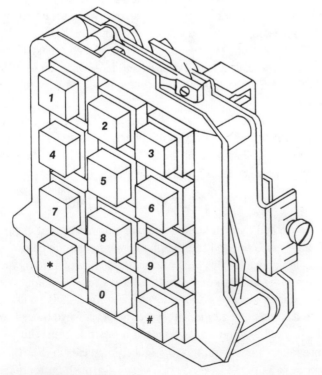

Fig. 1-15. A typical tone dial.

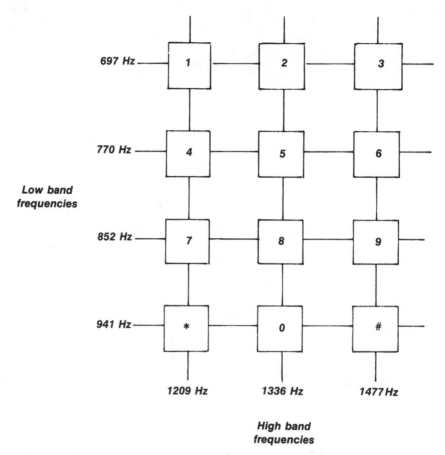

Fig. 1-16. Frequencies generated by a tone dial.

Pressing a button results in the generation of two tones: a low-band, and a high-band tone.

Pressing the number 3 button causes a low-bandtone of 697 Hz and a high-bandtone of 1477 Hz to be generated. These two tones will be converted by the central office, using special electronic filters, as being the number 3.

Generating the frequencies can be accomplished by an inductor-capacitor (LC 32opg type) resonant circuit, or by a master clock run by a crystal oscillator divided by a special integrated circuit (IC) to the needed push-button tones (42opg type).

## PROJECT INSTALLATION

All enhancement projects presented within this book must somehow be connected to the telephone line for proper operation. The most logical place for this connection is the junction box. Figure 1-17 is an illustration of the two

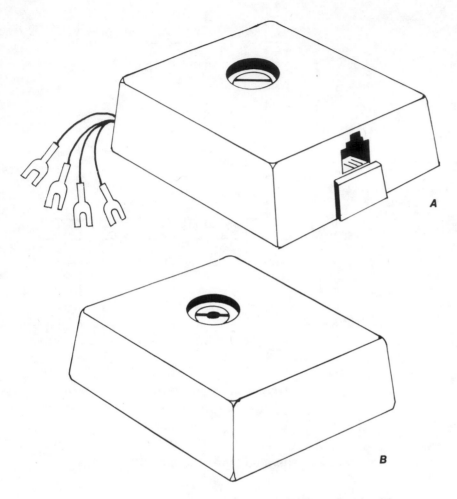

Fig. 1-17. Junction boxes (A) modular box. (B) the standard box.

most widely available junction boxed (or connection block) used in the installation of a home telephone.

Figure 1-17A is the junction box used with the newer type of telephone sets being manufactured today, this is the *modular* type. To install a telephone using this type of connector, all you must do is to snap a plastic line-cord terminator into the junction box. Four small piano wire conductors make contact with the plastic terminator allowing an electrical current to flow between the two points. At this time, the telephone is connected to the line.

You might have noticed that telephone coil cords also use this type of connection. If your telephone makes use of these plastic terminators, your telephone is said to be of the full modular type.

Presented at Fig. 1-17B is the older type of junction box. To connect a telephone to this type of installation, unscrew the plastic cover to expose four

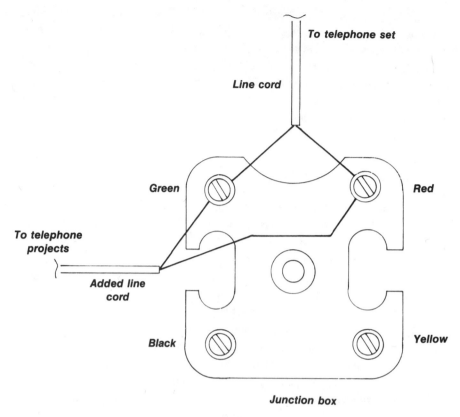

Fig. 1-18. How to connect telephone projects to a junction box.

screws that have the basic line-cord colors fastened to them (green, red, yellow, black). Figure 1-18 shows how these wires are fastened to the telephone set. All you must do is to match the wire color of the telephone line-cord to the printed wording located on the junction box surface. An example of this is to connect the green wire of the telephone line-cord to the screw marked "GN". If your telephone is of the older type, remember to also fasten the yellow line-cord wire to the "GN" junction box terminal. Then connect the red line-cord wire to the "RD" terminal.

Having this basic installation concept in mind, you can use this procedure to make a permanent connection to the junction box for any one of the enhancement projects.

## CONNECTION TO THE MODULAR JUNCTION BOX

Before you do any wiring, you must have on hand a telephone line-cord using a module plug (purchase this type of line-cord only if the junction box in your home is the same represented in part "A" of Fig. 1-17) on one end, while the other is terminated with what is called *spade lugs*. This cord is said to be a ¼ modular line-cord. On the end of the cord with the spade lugs, cut off the yellow and the black wires. These wires are not used, and should be removed to prevent accidental shorting. For final installation, just fasten the modular plug into the junction box, making sure the green and red wires on the opposite end of the cord are not shorted.

## CONNECTION TO A STANDARD JUNCTION BOX

If you have the older type of telephone installation, you can fasten a second line-cord to the box, but you must first remove the plastic cover. Do this by loosening the screw located at the center of the box. When the cover's removed, you will be confronted with the wiring presented in Fig. 1-18.

Using a line-cord terminated with spade lugs at both ends, cut off the yellow and black wire leads on *both* ends. These two wires will not be used by any of the telephone projects. Now connect the green and the red wires of the line-cord to their corresponding screw terminals (GN / RD), making sure the green and red wires at the opposite end are not shorted.

Due to the tremendous growth of the telephone industry brought about by the recent FCC ruling, telephone parts, like line-cords, are available in hardware and department stores. You can also purchase parts in one of many telephone specialty stores that have popped up across the country. The only thing to keep in mind is to shop around for the best possible price. Practically all telephone parts (line and coil cords) are imported. I find that imported telephone parts are pretty much reliable under normal usage, but the pricing from one store to another varies greatly, so look for sales.

So far, I have introduced you to the basic parts that make up a standard telephone: the ringer, tone and rotary dials, handset, and the telephone electronics. How these various components are connected together to allow the user to access a central office is the topic of this section. The diagram presented in Fig. 1-19 is what is called an *exploded diagram* of a home telephone. From this diagram, you can see how the various components are mounted on the telephone baseplate. The handset and the line-cord are clipped in place, while the rotary or tone dial is screwed to the mounting brackets. The ringer is screwed to the base while the hookswitch and the network are riveted in place. Let's stop here for a moment. Previously, I mentioned a network (not to be confused with the telephone switching equipment). The telephone network is where the electronics presented in Fig. 1-8 are located. On older models, this network is in the form of a metal box where the connections to the various points of the electronic circuit are made by screwing down the wire leads of the associated

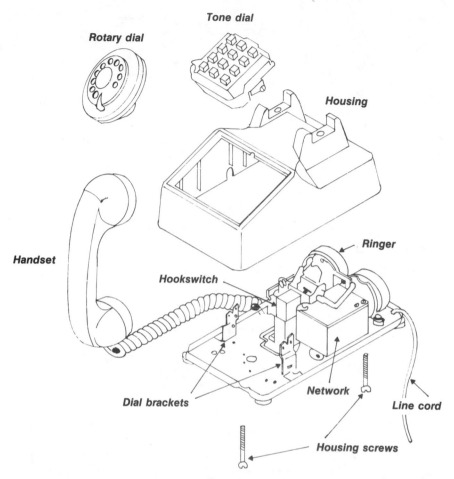

Fig. 1-19. Exploded view of a standard desk-type telephone.

component parts. The network of a telephone is like an engine block to a motor. All components are somehow connected to this one central point.

Figure 1-20 is an illustration of the connector locations of the older box-type telephone network. Each connection point is labeled with a letter, for example: R, GN, B, etc. Each letter corresponds to a different location within the electrical circuit of the telephone.

Another widely used network is illustrated in Fig. 1-21. This network uses a printed circuit board (PC board) that contains special push-on connectors in place of the screw terminals that are found on the older box assembly.

The termination lettering is identical for both networks, except that the PC board assembly contains an additional two connection points (E1 and E2). These termination points are used as a common termination point and have no connection to any of the telephone electronics.

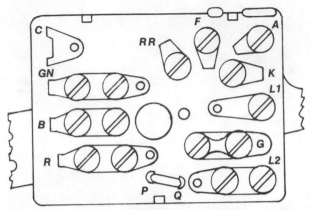

Fig. 1-20. Termination points on the old style metal box network. **(Courtesy of Corinth Telecommunications Corp.)**

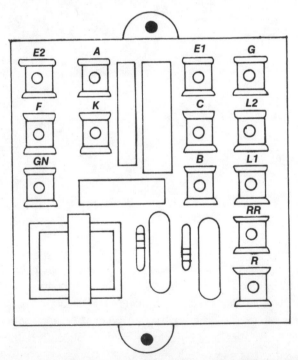

Fig. 1-21. Termination points on the newer type PC board telephone network. **(Courtesy of Corinth Telecommunications Corp.)**

Due to the construction of the box-type network this reliable device is used so that no operator contact can be made to the electronic components, but it is a different case with the PC board network. It is recommended that while experimenting with your telephone you exercise extreme caution. The most

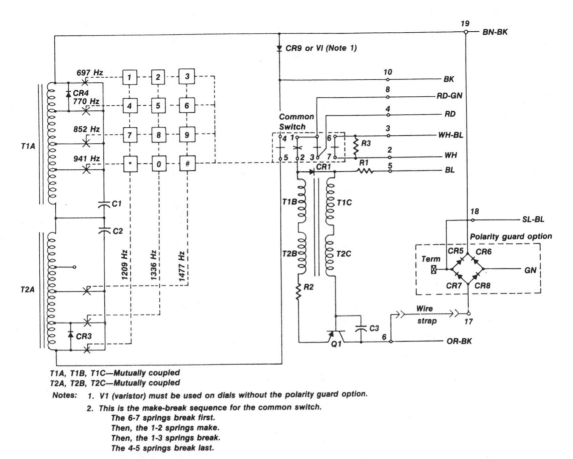

Fig. 1-22. Electronic Schematic on the older type 32G-type tone dial. **(Courtesy of Corinth Telecommunications Corp.)**

sensitive component on the board is the induction coil. The fine wires of the coils are exposed, and can be easily broken with a careless movement of a screwdriver. Again, exercise caution.

The last component that will be covered in this chapter is the tone dial. Presented in Fig. 1-22 is the schematic diagram of an older 32opg tone dial, while Fig. 1-23 is the schematic of the newer 42opg dial. Both dials have what is called a *polarity guard,* CR5 to CR8 in the 32opg, and CR4 to CR7 for the 42 dial. This bridge rectifier allows the tip and ring wires of the telephone line-cord to be reversed, but still allowing the proper voltage polarity needed by the electronic circuit.

Another point to mention is the 32opg tone dial uses the LC (coil/capacitor - T1 and T2) tuned circuit to drive transistor Q1 into oscillation at the required frequencies needed for tone dialing. The 42opg uses the newer technology of digital electronics to generate the required tones.

**22**

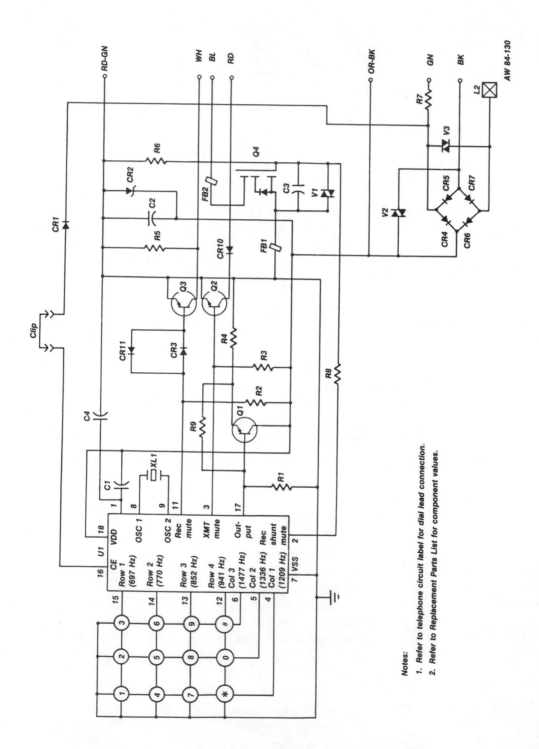

Fig. 1-23. Electronic schematic of the frequency synthesized tone dialer. **(Courtesy of Corinth Telecommunications Corp.)**

Notes:

1. *Refer to telephone circuit label for dial lead connection.*
2. *Refer to Replacement Parts List for component values.*

**23**

**186145-101**

**2500 Type Telephone Circuit (44 M Non Modular)**

NOTES:

1. BIASED RINGER CUT-OFF CONTROL BY CUSTOMER: BEND STOP NEXT TO DETENT ON VOLUME CONTROL SO THAT IT CLEARS RIM OF RINGER FRAME. THIS PROVIDES EXTRA CONTROL POSITION IN WHICH RINGER ARMATURE IS LOCKED.

2. WHEN USED WITH K1A1 OR K1A2 KEY TELEPHONE SYSTEM, BLACK CONDUCTOR OF MOUNTING CORD IS THE "A1" LEAD AND YELLOW IS THE "A" LEAD.

3. TO PERMANENTLY SILENCE RINGER: TRANSFER BLACK RINGER LEAD TO "A" TERMINAL ON NETWORK.

4. BROWN LEAD CONNECTED TO DIAL STATIC SHIELD.

5. TO DISABLE THE POLARITY GUARD FEATURE RECONFIGURE THE DIAL AS FOLLOWS:
   A. REMOVE THE OPTION CLIP FROM THE STORAGE (LOWER) NOTCH ON THE CIRCUIT BOARD AT THE REAR OF THE DIAL.
   B. PLACE CLIP IN THE POLARITY GUARD DISABLE NOTCH (UPPER).

6. * CONDUCTORS TAPED AND STORED.

ISSUE NO. —

BIAS SPRING ADJUSTMENT
HIGH BIAS POSITION
LOW BIAS POSITION

130 (BA)470 RINGER IS SHIPPED FROM THE FACTORY WITH THE BIAS SPRING IN THE HIGH BIAS POSITION. THE RINGER IS ADJUSTED TO RING AT 77 VOLTS AT 20 HZ. IN THE HIGH BIAS POSITION.

FOR LOWER VOLTAGES AND 30 HZ. RINGING, THE BIAS SPRING MAY REQUIRE MOVING TO THE LOW BIAS POSITION.

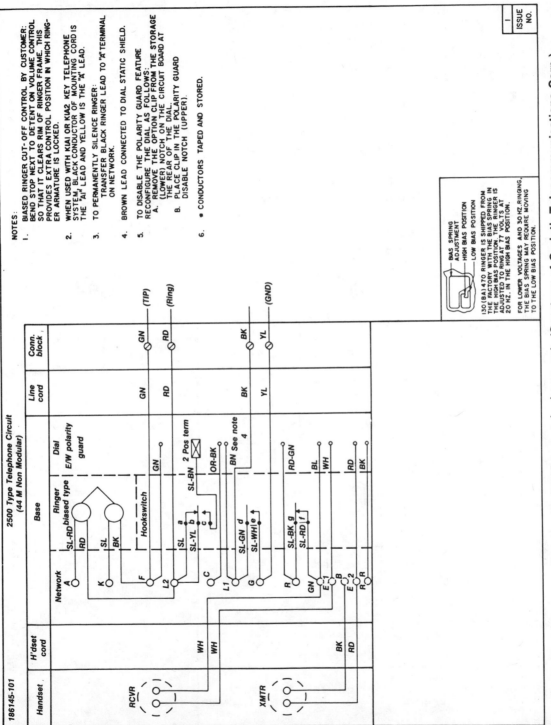

Fig. 1-24. Wiring diagram of a standard 500-type telephone set. (Courtesy of Corinth Telecommunications Corp.)

24

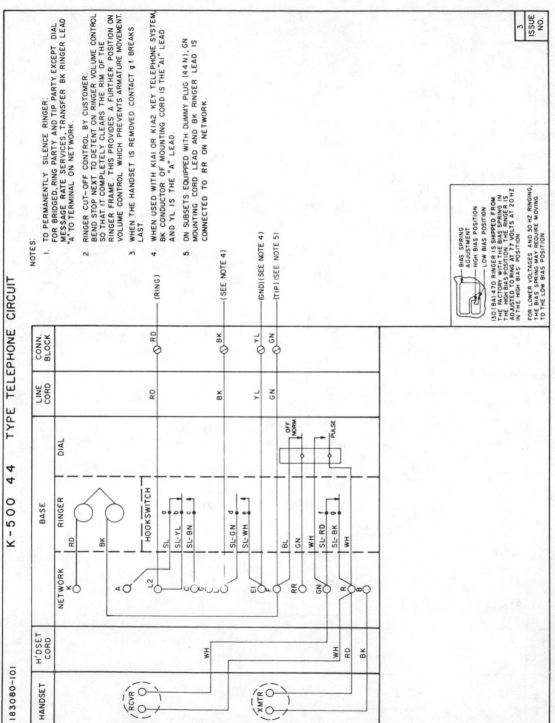

Fig. 1-25. Wiring diagram of a standard 2500-type telephone set. (Courtesy of Corinth Telecommunications Corp.)

By referring to the schematic of the 42opg dial, U1 is the heart of this type of frequency synthesis. When one of the available twelve buttons is pressed, one row and one column pin of the IC are shorted together, producing the corresponding digital frequency at the output located at pin 17.

## TELEPHONE WIRING

Through the courtesy of the Corinth Telecommunication Corporation, Figs. 1-24 and 1-25 show the wiring diagrams of the standard 500 (single line rotary dial telephone) and the 2500 (single line tone dial telephone). These diagrams are presented to illustrate the simplicity of a telephone instrument and to help the experimenter reassemble their telephone if they got carried away removing wires.

# 2 Assembly Tips

SINCE ALL THE PROJECTS PRESENTED IN THIS BOOK USE 12 VOLTS OR LESS IN their operation, the assembly of any project is completely safe and straightforward. This chapter includes some basic electronic assembly information that you will find extremely useful, not only for the enclosed enhancement circuits but for all other future projects.

This book was written with the novice electronic hobbyist in mind, but even if you have more experience, you should find it helpful to review this chapter, particularly the section on how to solder.

## COMPONENT SELECTION

The components used in the enhancement projects include integrated circuits (ICs), transistors, resistors, capacitors, and diodes. If you have never built an electronic project before, these components might seem strange and hard to understand, but when you use them within circuits, both published, and those of your own design, you will gain the understanding of how to put a large assortment of components to work properly.

It is the scope of this chapter to introduce the beginner or the novice on how to intelligently select and use basic electronic components.

# RESISTORS

When you open the case of any electronic device, you will come across resistors, which are common components used with most modern electronics. As the name implies, resistors literally *resist* the flow of electrical current. Resistors are used to limit the current flow of a circuit that might otherwise burn up. When you choose the correct resistor value, the resistor can supply a trickle of current to transistor, to adjust the operating characteristics, allowing the transistor to operate effectively within the manufacturers' allowable current range.

Most resistors are made of carbon and are available in a wide range of values. These values range from one to several million ohms. An ohm is a unit of resistance. To describe a resistor having a value of 50 ohms is to say that this component has 50 units of resistance. Three or more color bands around one

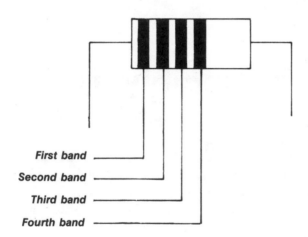

| Color | First Band | Second Band | Third Band | (Multiplier) |
|---|---|---|---|---|
| Black | 0 | 0 | $10^0$ | 1 |
| Brown | 1 | 1 | $10^1$ | 10 |
| Red | 2 | 2 | $10^2$ | 100 |
| Orange | 3 | 3 | $10^3$ | 1000 (Kilo) |
| Yellow | 4 | 4 | $10^4$ | 10000 |
| Green | 5 | 5 | $10^5$ | 100000 |
| Blue | 6 | 6 | $10^6$ | 1000000 |
| Violet | 7 | 7 | $10^7$ | 10000000 |
| Gray | 8 | 8 | $10^8$ | 100000000 |
| White | 9 | 9 | | |
| Tolerance band (fourth band) | | | | |
| | | Gold | 5% | |
| | | Silver | 10% | |
| | | None | 20% | |

Fig. 2-1. Resistor color code.

end of the resistor are used to denote the resistance in ohms. This resistor color code can be seen in Fig. 2-1. To see how the code works, let's determine the value of a resistor with yellow, violet, and red bands. Always read the color bands beginning with the band closest to the end of the resistor. Yellow, being the first of the three color bands, has a corresponding value of four. Violet is the second band, and Fig. 2-1 shows that violet has a corresponding value of seven. The third band determines the factor of ten, by which the first two numbers must be multiplied to find the total resistance presented by the bands. Red corresponds to a value of $10^3$, or a multiplication factor of 100, so the sample resistor is 47 × 100 or 4,700 ohms.

Often on a printed electronic schematic, you will see a resistor value followed by a *k* (kilo) or *M* (Meg). The use of the k, indicates that the resistor value given should be multiplied by 1,000. The M indicates that the value should be multiplied by 1 million. For example, a 47 k resistor has a value of 47,000 ohms and a 1.5 M has a value of 1.5 million ohms. Many resistors have a fourth band that specifies the component's tolerance. A *gold* band means the color code value is within 5 percent of the actual value and a *silver* band means the color code value is within 10 percent of the actual value. The resistor has a 20 percent tolerance if there is no fourth band.

Resistors also come in assorted Watt ranges: ¼ W, ½ W, 1 W, 2 W, 5, 10, 25, etc. *Watts* basically indicate the amount of current versus the resistance that can flow within the resistor without causing damage to the component. This damage is in the form of heat. If a specified Wattage is exceeded, the resistor can actually begin to smoke. The Wattage of a circuit can be determined by any one of the following calculations:

$$\text{Wattage} = \text{Total Current} \times \text{Total Voltage}$$
$$\text{or} = \text{Total Current}^2 \times \text{Total Resistance}$$
$$\text{or} = \text{Total Voltage}^2/\text{Total Resistance}$$

All the projects that will be presented in this book require the use of very low voltage. All resistors used have a rating of ¼ Watt, but if ½ W sizes are available, by all means, use them. There will be no adverse effect on circuit operation. The tolerance of the resistors used can be 10 to even 20 percent. Again, this will have no effect on the operation.

## POTENTIOMETERS

A *potentiometer* (or pot) is an adjustable resistor, and therefore quite useful in circuits that require frequent adjustments. An example of a potentiometer that you should be familiar with is the tone/bass/volume control of your AM/FM radio. Figure 2-2 is an example of a standard potentiometer used as a volume control. This is by no means the only configuration that potentiometers are available in, for additional examples, just thumb through any electronics mail order catalog. You might be very surprised.

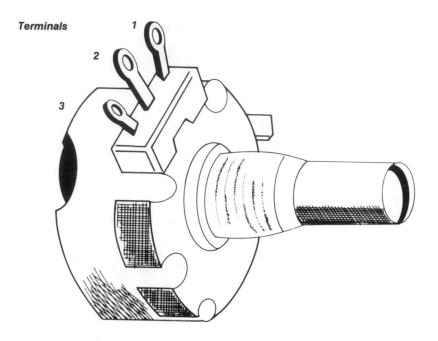

**Terminals**

1

2

3

Fig. 2-2. A standard potentiometer (variable resistor).

In the AM/FM radio example, if the position of the knob on the volume control is changed, the amount or intensity of sound that is delivered to the speaker also varies in proportion to the setting. When turning the knob, the internal resistance of the potentiometer changes. By applying a sound signal (also called an *audio signal*) to a terminal on the potentiometer and by varying the resistance, the intensity of the sound changes. For a better understanding of this, refer to Fig. 2-3A. With the potentiometer set at its low-resistance position, the audio signal has very little opposition (resistance) to flow of current, but when adjusted to its high-resistance position (Fig. 2-3B), the flow of the audio signal encounters a greater opposition. This opposition will impede the flow of all (or most) of the audio signal, making the sound available to the speaker very low.

Much like fixed value resistors, potentiometers come in a wide variety of styles, resistance values, and Wattage ratings. However, potentiometers have no color code to indicate the resistance value. These values are usually stamped directly on their bodies.

## CAPACITORS

*Capacitors* are components that have the ability to store an electrical charge. The amount of charge depends on the size of the internal capacitor plates. Capacitor values range from thousands of microfarads (large) to hundreds of picofarads (small). Using this ability to store an electrical charge, the capacitor

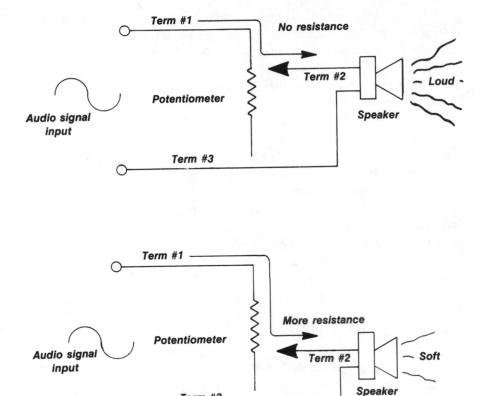

Fig. 2-3. The effects of a potentiometer on an audio signal.

acts as a reservoir of electrons to provide a current (electrical flow) during brief voltage interruptions. An example is a power supply. When a small dc voltage is present, the capacitor stores a charge. But when this available dc voltage is missing, the capacitor discharges. Thus providing a voltage when there is none. The charging and discharging of this capacitor (also known as a *filter capacitor*) converts what is known as a pulsating dc voltage to a pure dc voltage.

This pure dc voltage can be compared to the voltage of a 9-volt battery used to power your transistor radio. The value of filter capacitors are very large—usually on the order of 1000 to 2000 microfarads.

Like Meg (M) and kilo (k) ohms used to indicate the large values of resistance, farads are used to indicate a unit of capacitance. A Farad is too large for practical work, however, so capacitors are measured in microfarads and picofarads. Microfarad means that the value of the capacitor is 1 times $10^{-6}$, or one-millionth of a Farad. In other words, a capacitor with a value of 2000 microfarads is equal to a capacitor with the value of .002 farads (i.e., 2000 × .000001 = .002 farads).

The symbol that indicates the microfarad is "$\mu$F", (2000 $\mu$F capacitor). To indicate even smaller capacitor values, the picofarad is used. Picofarad is one-millionth of a microfarad, or 1 times $10^{-12}$.

If the value of a capacitor is 100 picofarads, by taking the numeral 100 and multiplying it by $10^{-12}$ the calculated value of this capacitor is .0000000001 farads. So it is not necessary to write all those zeros, the "pF" designation is used, it means picofarads.

Capacitors are ideal for sound (audio) circuits. They pass ac signals while stopping a dc voltage. Yes, an audio signal is considered an ac voltage. The capacitor value usually used in audio circuits is .1 $\mu$F to .01 $\mu$F. Like resistors, capacitors come in a wide range of tolerances. Frequently, the actual value of a capacitor will range from 20 to even 100 percent of the value printed on the device. This means capacitor substitutions are permissible in *almost* all electronic circuits. For example, a .025 $\mu$f capacitor can almost always be substituted for a .02 $\mu$f or even .01 $\mu$f unit with no difficulty. In *tuned circuits* like radios or rf oscillators, however, changing the capacitor value from what is indicated can adversely affect the operation of the circuit. If this substitution were made in a television transmitter, the FCC would be knocking at your door and present you with a hefty fine.

## MAKING A CAPACITOR

Presented in Fig. 2-4A is a graphic representation of the internal construction of a capacitor. A capacitor requires two electrically conductive plates placed next to each other, but not touching. To produce capacitors of different values, the manufacturer varies the physical size of, and the spacing between, the plates. For smaller value capacitors, a plastic sheet is placed between the plates. This sheet is called the *dielectric*.

To manufacture larger value capacitors (filter capacitors), several layers of plates are connected in parallel (see Fig. 2-4B). This capacitor assembly, which now resembles a sandwich, is tightly rolled into a cylinder, coated with a protective layer of plastic or wax, then sealed inside a container (usually aluminum) with wire leads protruding from each side (an *axial type*) or from the bottom (a *radial type*).

## DIODES

The diode is one of the simplest semiconductor components. All modern diodes are said to be pn junction types. A pn junction diode consists of two small plates of silicon, selenium, or germanium to which impurities have been added to form either an n-type or p-type material. These two plates (one being n-type material and the second being p-type) are then physically connected. When connected, some of the extra electrons in the n-material, which has a negative charge, move to the p-type material, which has a positive charge, and some

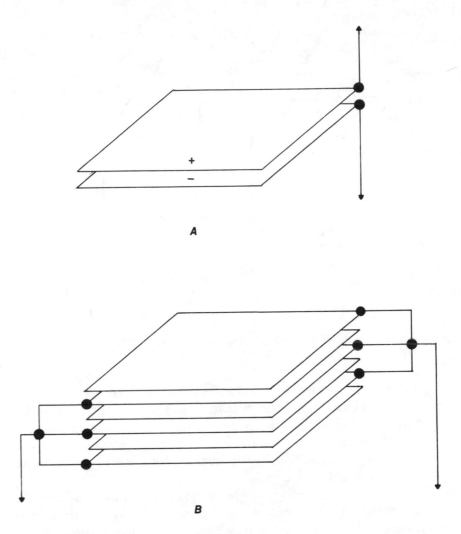

Fig. 2-4. Plates of a capacitor and how they are wired together.

from the p move to the n-material. This process occurs because a silicon plate is considered to have a positive charge, but it still contains a small amount of extra negative electrons. In reverse, a silicon plate can be considered negative but it will still contain extra positive electrons. When the p and n materials are connected together, equal amounts of electrons cross between the n to the p material and from the p to the n material. The region of opposite charge that is formed around this contact area is called the *depletion layer*.

To *bias a diode* is to pass an electrical current in either the "forward" or "reverse" direction. The forward bias condition can be seen in Fig. 2-5A. There the polarity of the voltage source (V) is connected in such a way that the positive voltage is applied to the p (positive) material and the negative is applied to the

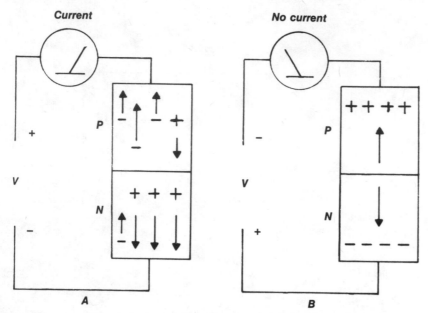

Fig. 2-5. The effects on current when forward and reverse biasing a pn junction.

n (negative) material. Because like charges repel, the charge carriers in both the p and n material are driven away from the power supply terminals and toward the junction. Using a power supply with a high enough voltage, the depletion layer will disappear, allowing the positive and negative charges to get closer together, until the opposite charges attract each other across the junction, thus allowing an electrical current to flow.

The reverse bias condition can be seen in Fig. 2-5B. In this case, the negative terminal of the power supply is connected to the p type material, while the positive terminal is connected to the n material. Because opposite charges attract, the positive electrons in the p material are attracted to the negative side of the power supply, while the negative electrons in the n-type material are attracted to the positive side of the supply voltage. This action widens the depletion layer of the junction and no electrons are allowed to flow through the junction, thus preventing any electrical current.

This diode theory is presented as an ideal condition, but in life, nothing is ideal. Even if a diode is reverse biased, there will still be a slight amount of current flow between the junction. This flow is called the *leakage current,* or simply *leakage.* By using a power supply, we can see the relationship between forward- and reverse-biased diodes. An ac voltage is made up of a positive and negative cycle. The signal will complete one cycle every 1/60th of a second, or 60 cycles in one second. With a diode placed within this voltage cycle, the diode will allow current to flow only when it is forward biased. When the negative

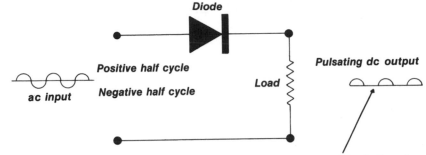

Fig. 2-6. The effects on an ac signal when rectified by a diode.

half of the ac voltage is present, the diode will be turned off (reverse biased) so no voltage will be allowed through.

The output of this diode is shown in Fig. 2-6. This circuit is called a *half-wave rectifier*. This type of rectifier is used to convert an ac signal to what is called a *pulsating dc*. To smooth out this pulsating dc voltage, a 1000 μF capacitor can be installed. Other types of rectifiers used in electronics are the full-wave (Fig. 2-7A) and the bridge (Fig. 2-7B).

## LIGHT-EMITTING DIODES

A very interesting component is the light-emitting diode (LED). The LED, when reverse biased, acts like a standard diode, but when forward biased emits a beam of light. Some LEDs emit up to hundreds of milliwatts of invisible infrared radiation. These LEDs are used in wireless remote controls in televisions and video cassette recorders. Other, less efficient, LEDs emit visible green, yellow, amber, or red light. The red LEDs are the easiest to manufacture so they are the most economical for the hobbyist.

Because of the low internal resistance of a standard LED, a resistor must be connected in series to limit the current drawn by the device, otherwise the LED will burn up. This resistor can be in the order of 1000 ohms (1k ohms ¼ Watt) for an applied voltage of 12 volts dc.

When soldering an LED, use a *heat sink* (alligator clip) to prevent excess heat from damaging the component, or you will be out $.20 for the LED. Figure 2-8 is a graphic representation of a standard LED.

## TRANSISTORS

A *transistor* is a semiconductor device that closely resembles two diodes connected back to back. The result is a device with two pn junctions. Diodes are simple pn devices. Transistors are pnp or npn semiconductors.

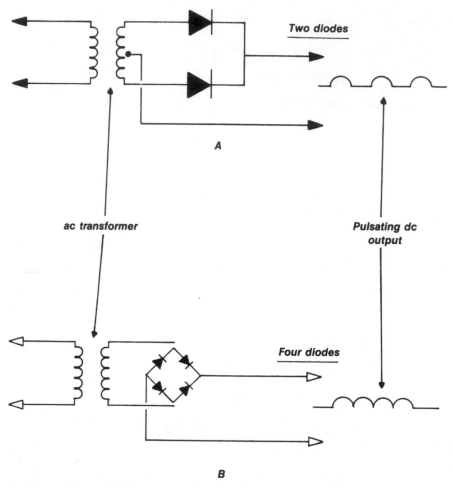

**Two diodes**

*A*

*ac transformer*

**Pulsating dc output**

**Four diodes**

*B*

Fig. 2-7. Typical rectifier circuits (A) shows a full wave rectifier (B) a bridge rectifier.

The center section of a transistor (the n section of a pnp or the p section of an npn transistor) has the ability to control current flow through the device, just like the grid of a vacuum tube can control electron flow. With no current connection to the center section, current cannot flow between the outer halves and the transistor is considered to be off. When a small amount of current is allowed to flow through the center section, the transistor is said to be turned on; current now flows between the two outer halves.

Because a transistor uses very small amounts of current to control a larger amount, the transistor is ideally suited as an amplifier. If a very large input signal is applied to the transistor, the device can be turned completely off. In this state, a standard transistor can act much like a switch.

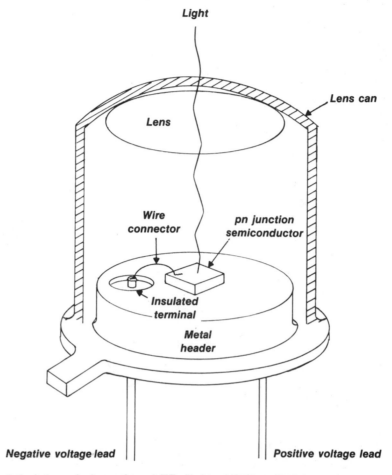

Fig. 2-8. Internal view of an LED (light emitting diode).

The reason for having two types of transistors (pnp/npn) is very simple. Unlike vacuum tube circuits, which allow the plate of the tube to be connected only to positive voltage, the transistor can be connected to positive or negative voltages. For example, the grounding voltage in a car made in the United States is negative voltage, while the grounding voltage for many Japanese products is positive.

For a negative ground circuit, you must use the npn transistor. For positive ground circuits, you must use the pnp. If you substitute a pnp for a npn transistor, the junction of the transistor will be reverse biased, causing little or no current to flow. Your project will not operate.

You can distinguish the difference between a npn and a pnp transistor by looking at the direction the arrow is pointing within the transistor symbol on a schematic. If the arrow points away from the circle, the transistor is a npn, if it points toward the circle, the transistor is of the pnp type.

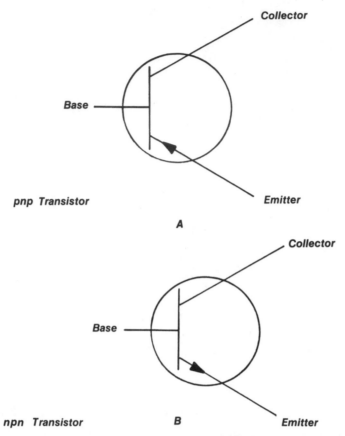

Fig. 2-9. Transistor schematic symbols. (A) shows a pnp transistor. (B) a npn transistor.

The three wire leads of a transistor also have names. They are the collector, base, and emitter. The arrow that distinguishes between the npn and pnp transistors also helps determine the names of the leads. The arrow is drawn on the leg of the transistor, called the emitter, while the center section is the base, and the third lead is the collector. (See Fig. 2-9.)

There are many types of transistors and circuit connections used to construct amplifiers (common emitter, common base, common collector), switching circuits, and oscillators. There is not enough room to explain them all. If you wish additional information on transistors, TAB BOOKS has a number of other fine publications that will instruct you in the theory, and operation, as well as provide you with the project diagrams, of transistor circuits.

## INTEGRATED CIRCUITS

*Integrated circuits* are highly condensed packages of electronic circuits. Most ICs include hundreds of transistors, resistors, and diodes, all packaged on a

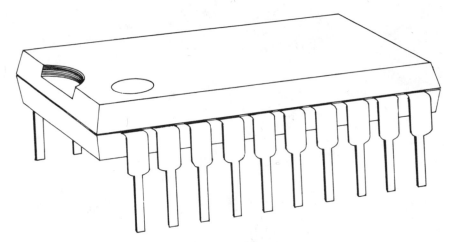

Fig. 2-10. A typical integrated circuit (IC). Note the dot over the #1 pin of the IC.

silicon chip that can pass through the eye of a sewing needle. Hundreds of ICs are available to the hobbyist at very reasonable cost. ICs fall into two broad categories: digital and linear. *Digital ICs* are always found in computer hardware because the output voltages of digital ICs are either positive or negative. Other ICs are called *tri-state*. The output from these ICs are positive, ground, or high impedance (or an open circuit).

The ICs used in many of the projects in this book are of the linear type. These ICs are working circuits in themselves and require only the addition of a small number of components for a complete project. A typical IC is illustrated in Fig. 2-10. This type is known to be a *dual inline package.* Other packaging is also available. Linear ICs are the best thing to come along in a long time especially where the hobbyist is concerned. Take the LM386 IC: except for the addition of five components, you can build a complete high-quality audio amplifier. In the days before integrated circuits, this amplifier took about six transistors and a large handful of resistors and capacitors, not to mention the time spent designing and calculating the component values. Linear ICs will save you both time and money on your next project, so make use of them.

# 3 Reading Electronic Schematics

SIMPLE SYMBOLS HAVE BEEN DEVELOPED TO HELP CIRCUIT DESIGNERS DRAW A diagram that represents a working project. This diagram is called a *schematic*. Figure 3-1 shows most of the circuit symbols used in this book but in the world of electronics, this number can run in the thousands. The number of symbols that you will be required to understand depends on the field of electronics you are entering. Whichever field you decide to go into, you will be required to read and understand schematics.

Figure 3-2 is a schematic of a circuit that you are very familiar with, a flashlight. Component #1 is the standard electronic symbol for a battery. Note that a battery is represented by the drawing of parallel lines. One set being shorter than the second. The drawing of a short line indicates that this is the *negative* side of the battery. In electronics, the polarity of batteries as well as some components must be observed or you run the risk of destroying voltage sensitive devices. Note the line that connects the positive side of the battery to the component labeled #2. This is the light bulb. With current passing through the bulb, the filament begins to glow. Energy will then be dispersed as visible light radiation. For the electrical current to continue its journey a connection must be made from the bulb to the next component. This is the on/off switch (component #3). With the switch's contact in the position shown, electrical current is blocked preventing the continuous flow of electrons from the positive

42

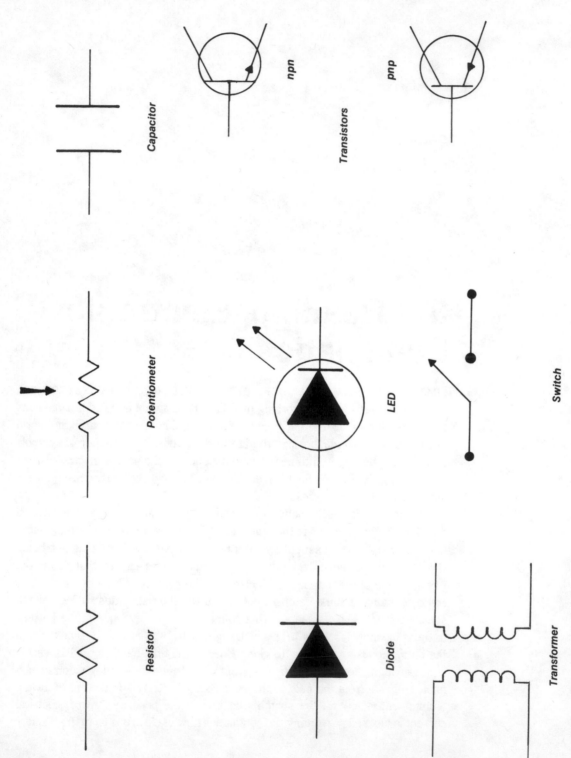

Fig. 3-1. A small sample of electronic schematic symbols.

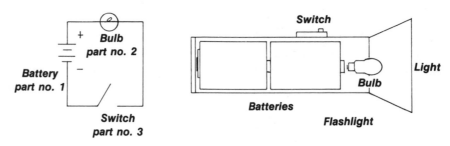

Fig. 3-2. A flashlight is a simple series circuit containing a battery, switch and a bulb.

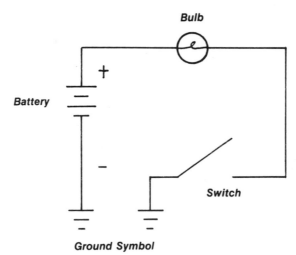

Fig. 3-3. Adding a ground symbol to the flashlight schematic.

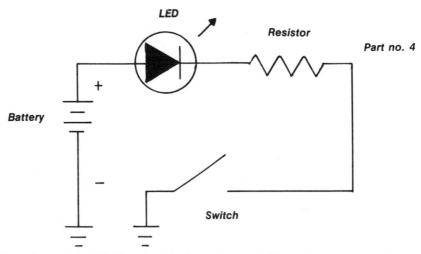

Fig. 3-4. Changing the flashlight bulb to an LED. Remember to add the resistor.

side of the battery to the negative. This can be compared with a drawbridge spanning a river. With the bridge open, cars cannot be driven across. But if the bridge were to close, traffic will then be allowed to continue to the other side. If the switch were to close, as with the bridge, electrons are then allowed to flow through its electrically conductive contacts to complete the journey to the negative side of the battery.

Figure 3-3 is the same flashlight drawing but with a slight difference. The line connecting the switch to the negative side of the battery has been replaced with a symbol that represents ground. In practice, all ground connections shown on a schematic are physically connected together. This is known as *common ground* or common return.

Figure 3-4 shows that the bulb has been replaced with another light emitting device, the *light emitting diode* or LED for short. Now that an LED is placed in the circuit, a current limiting resistor has to be added. This resistor limits the total current to the LED, without it the LED will burn out. This resistor is component #4 in the schematic.

# 4 Telephone Project Construction

SOME OF THE CIRCUITS IN THIS BOOK USE SO FEW COMPONENTS THAT THE making of a PCB (printed circuit board) is not really necessary. However, operation of both simple, and the more complex circuits listed are considerably enhanced by using PCB's. For the more simple projects, you can use grid perforated board available from any electronic outlet.

Perforated boards are ideally suited for small electronic projects as well as for experimental circuits, since modifications and component changes can be made quickly and easily. Also, since all component leads are exposed it is easy to check the operation of the circuit with a multimeter.

For more permanent projects, you might want to take advantage of the artwork presented for a selected number of projects. This artwork can be found in the end of the chapters. Etched printed circuit boards are professional in appearance, sturdier, and more reliable than perforated board construction, for this reason, you might consider making your own, or send the enclosed artwork to a manufacturer so he can etch the PC board for you at a minimum cost.

The easiest way to a finished product is to send the artwork of the project along with a diagram indicating the size of the holes to be drilled, to a PCB manufacturer. Purchasing a manufactured PCB is just like buying a car. Shop around for the best possible price. Manufacturers charge their customers by the total area of the board in square inches. This price can range from $0.20

to $0.50 a square inch. To find the cost of a board lets go back to elementary arithmetic.

PRICE per Board = WIDTH × LENGTH × Cost per sq. inch.

Be careful, some manufacturers also charge you an additional $0.01 to $0.02 for each hole drilled in the board. Other manufacturers charge nothing. Request price quotes from a number of firms before sending them the artwork. And when you find a company you like stick with them. You will have less problems with future purchases.

When you have found the company, sent the artwork, and indicated to them that the scale of the pattern is 1:1 and to drill all holes 0.018 in. This is the standard hole size used with ICs, ¼ watt resistors and small capacitors. Within two weeks the finished product should be delivered to your door. What could be easier?

Another method that can be used to make your own PCB's is the *photoresist method.* It is a professional method of fabricating and should not be considered by the beginner. This method uses a special printed circuit board with a light sensitive coating on the copper side. When light is passed through the PCB negative artwork (that is in direct contact with the copper) a chemical change takes place. After exposure to light for a predetermined length of time, the board is now immersed into a chemical solution that hardens the area of the PCB that electrical connections are to be made (this trace is called the *land*).

The board is placed in the etching solution where the unused copper is removed. This is called *etching.* After the board is rinsed in clean water and dried you can begin to drill the holes where needed. A drill with a carbon-tipped bit is strongly recommended. If any other drill bit is used you run the risk of lifting the delicate PCB pads and destroying the board. Complete kits using this method of PCB fabrication are available from large electronic houses for about $50.00. If this method suits your needs let's discuss what is involved to transfer your artwork to a finished board.

In order to produce a board directly from the enclosed artwork, you must inspect it under a strong light, looking for breaks or bridges in the land (*bridges*— when two or more land traces intersect at a point where electrical connections are *not* to be made). When you are satisfied that the artwork is clean, bring the artwork to a photographer and ask for a *negative* print of the page. This negative is just like a negative of a photograph. All areas that are clear will become black and all areas that are black will become clear.

Place the photographic negative on the sensitized copper board (wording side up) and expose it to a strong light source located 12 inches from the PCB. The amount of time needed for proper exposure is obtained by experimentation. The instruction booklet that comes with the photo etching kit will give you more

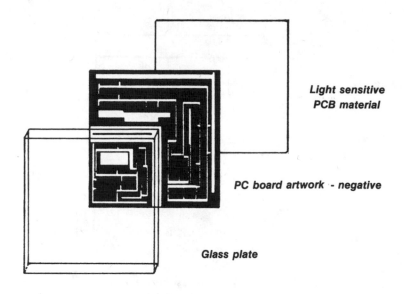

*Light sensitive
PCB material*

*PC board artwork - negative*

*Glass plate*

*Place all 3 materials in contact with each other and expose to strong light.*

Fig. 4-1. Typical way to expose light sensitive PC board material to UV light. All three materials (Glass, PCB, and PCB negative) are sandwiched together, ready for exposure.

precise timing. Refer to Figs. 4-1 and 4-2 to further explain the photoresist method of PC board fabrication. After the board is developed and dried you can remove the unwanted copper by using the etching solution. When etching is complete and the board is rinsed and dried, the material is now ready for drilling using a carbon-tipped drill bit, and component mounting.

## TAPE AND DOT PROCESS

The most commonly used method of fabricating your own PC board is the *tape and dot process.* This material uses pre-cut plastic dots and tape manufactured in a wide variety of sizes. These materials when placed on the copper side of a board, resist the etching solution. When etching is complete, these dots and tape can then be removed to reveal the copper layout of the board.

Tape and dot kits are available in 4X, 2X and 1X sizes (Refer to Fig. 4-3 for sample sizes of IC pads). But for direct etching of the enclosed artwork, 1X material must be purchased. To use this tape and dot process, you must transfer the printed artwork to the surface of the copper-clad board. A way to solve this problem is to use carbon paper. Before you begin to transfer the image, clean the copper surface with a steel-wool pad to remove any oxidation that might have been formed. If this is not done, any foreign substances on the board will prevent the etching liquid from dissolving the unwanted copper, rendering

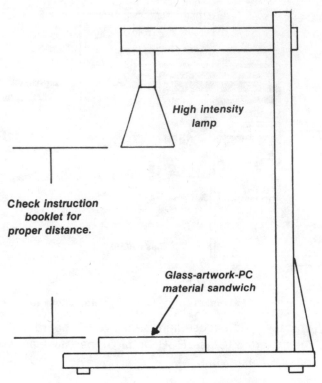

*High intensity lamp*

*Check instruction booklet for proper distance.*

*Glass-artwork-PC material sandwich*

Fig. 4-2. Recommended stand for exposing PCB to UV light.

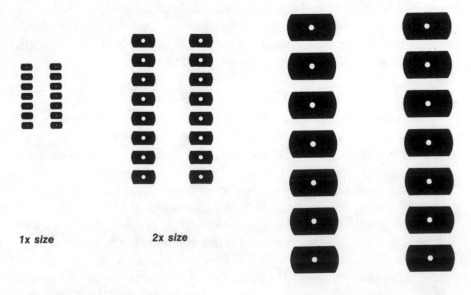

*1x size*

*2x size*

*4x size*

Fig. 4-3. PC board artwork scales.

the board useless. With this in mind, place the carbon paper (carbon side down) in contact with the copper facing of the board. Place the desired artwork over the carbon paper (see Fig. 4-4). If necessary, trim both the carbon paper and artwork to the size of the board. Using a ball-point pen or hard pencil, trace out the artwork. Don't use any more pressure than is needed, you can destroy the layout. It is only necessary to roughly trace out the needed pattern. When you have completed tracing the artwork, remove the layout and carbon paper, you should now see an outline of the desired copper trace on the board.

To remove the dots from their paper backing use an artist's knife and gently lift (see Fig. 4-5). With the dot still on the point of the knife, gently place the dot in position on the copper where the carbon paper left a trace. Repeat this procedure until all locations needing a hole drilled are covered by a dot.

Interconnect these dots with the tape. The thickness of the tape should be about 1/16 (0.062). When all needed traces are covered by either the dots or tape, it is now time to submerge the board in etching solution. This procedure will take about 10 to 15 minutes. But by heating the etching solution a bit, you can decrease this time by half. A method of heating the solution is to run hot water in a pail. With the etching solution in a glass jar, place the jar in the pail and leave it there for about 1/2 hour. The heat of the water will be transferred to the solution. Now the board can be etched in the usual manner. When etching is complete, rinse the board in cold water to remove any solution. Use the artists'

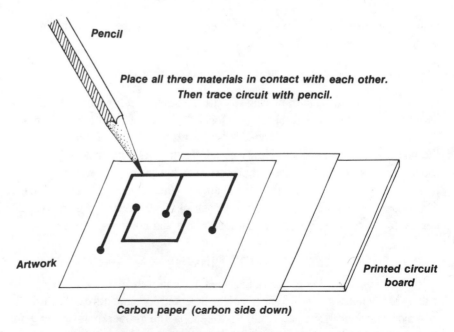

Fig. 4-4. Using carbon paper and pencil to trace out PC board artwork from book pages.

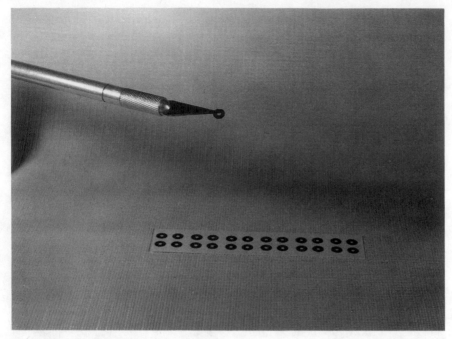

Fig. 4-5. Photo showing the dot an tape process to make PC boards.
**(Courtesy of Del-Phone Industries Inc.)**

knife to remove the tape and dots. The board is now ready for drilling and
component mounting.

## PEN AND INK PROCESS

Still another method of placing resist on a blank circuit board is to use the
*resist pen.* This pen is available at your local electronic store at a modest price.

Using the procedure described, trace the artwork on the copper side of
the board with carbon paper. Now instead of using the tape or dots, just draw
the interconnecting traces and dot pattern with the resist pen. Allow the resist
to dry then etch the board in the usual manner. When etched and rinsed in clean
water, use steel-wool to remove the resist ink. The artwork drawn with the
pen is now a finished board waiting for drilling and component mounting.

## DRILLING

When your board is etched and rinsed in clean water, the next step is the
drilling of the holes so the wire leads of the selected components can fit through.
A major consideration that must be taken, is the proper hole spacing for
components like integrated circuits, relays, and other devices that have leads
that cannot be bent to the desired position.

To help in the drilling, use a perforated board, the type used for project construction. With pre-drilling hole spacing of ⅒ inch, the board can be used as a template. Align the perf board on the PCB where the holes are to be placed, then using a small power drill, carefully drill holes through the perf board to the PCB. You will then have perfect hole spacing.

## Hole Sizes

For standard components like integrated circuits, small capacitors, and glass signal diodes, use a drill-bit with a diameter of .01 inch. For components using heavy wire leads, use a drill-bit with a diameter of .043 inch. Examples of components that use heavy leads are filter capacitors, and diodes used in rectifier circuits.

## COMPONENT MOUNTING

When mounting components on a finished PC board, neatness counts. Don't insert the components in such a way that the device is suspended in the air.

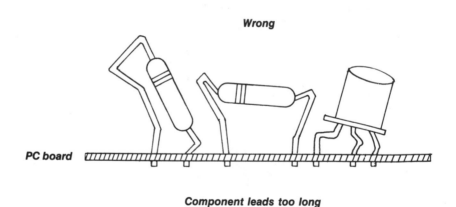

*Wrong*

*Component leads too long*

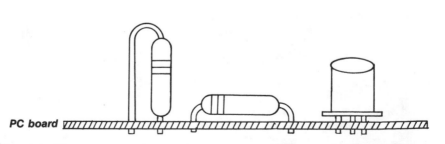

*Correct*

Fig. 4-6. The right and wrong way to insert electronic components on a PC board.

52

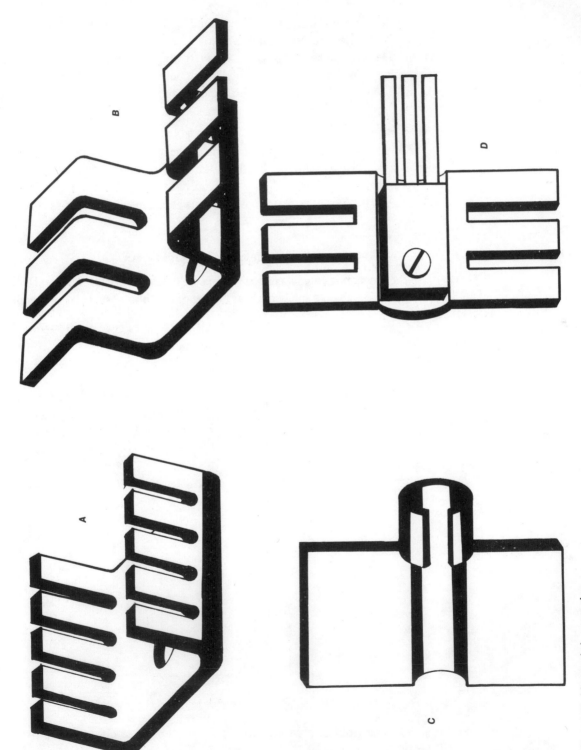

Fig. 4-7. Heat sink samples.

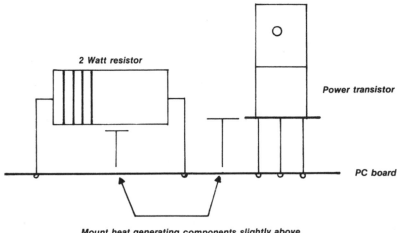

2 Watt resistor

Power transistor

PC board

Mount heat generating components slightly above
PC board to allow for natural cooling.

Fig. 4-8. The correct way to mount heat generating components.

All parts should be in physical contact with the PC board material. See Fig. 4-6. There is one exception to this statement. Power components such as power resistors or power transistors generate heat. This heat must be dissipated in the air, otherwise the operation of the component deteriorates.

Use a heat sink. *Heat sinks* are metal devices that resemble bird wings fastened to the power component. The heat generated by the part is transferred to the heat sink through a silicon compound placed between the two devices. With the heat now being transferred to the heat sink, the bird like wings dissipate the unwanted heat into the air. This heat dissipation keeps the operating temperature of the electronic component to an acceptable level. Heat sinks are available commercially through electronic mail-order houses, and they come in a variety of sizes and shapes to accommodate a large number of components. See Fig. 4-7 for examples of heat sinks.

You can also mount heat generating components in such a way that the body of the device is not in physical contact with the board, but don't exaggerate this spacing. Refer to Fig. 4-8 for an example of this.

When all components are mounted and soldered, check the board for any unwanted solder bridges, (copper traces are shorted together by using too much solder), component polarity, and proper values. When you are sure all is correct—and only then, apply the voltage to the project and check for proper operation.

# 5 Soldering Techniques

GOOD SOLDERING PRACTICES ARE ESSENTIAL FOR RELIABLE OPERATION OF ANY project. If you are an electronics novice or an experienced technician, please read the following carefully, and practice soldering scrap wire together before attempting to solder any components to a PC board.

If you don't own a soldering iron, you should purchase a "soldering pencil" with a maximum rating of 25 to 40 watts. Do not purchase a soldering gun. The high power associated with a gun can destroy delicate integrated circuits and other heat sensitive components with its high heating level.

When first used, the tip of the soldering pencil must be tinned with standard electrical solder. The word *tinning* is a term used by technicians that means, when the iron is first used, dab a small amount of solder on the tip when the iron reaches normal operating temperature. After a few seconds, using a damp cloth, wipe off the excess solder. After repeating this procedure a few times, you will notice that the irons' tip will become shiny.

Without tinning, a coat of oxidation will be formed on the tip. If this oxidation is not removed, soldering any component will be impossible because the solder will only bead-up on the tip, and there will be no transfer of heat from the iron to the component. When this happens, a *cold solder joint* will be made, or you might risk burning out the component.

Acid core solder should never be used for soldering electronic components, this is a corrosive material and will only damage electronic parts. A high quality

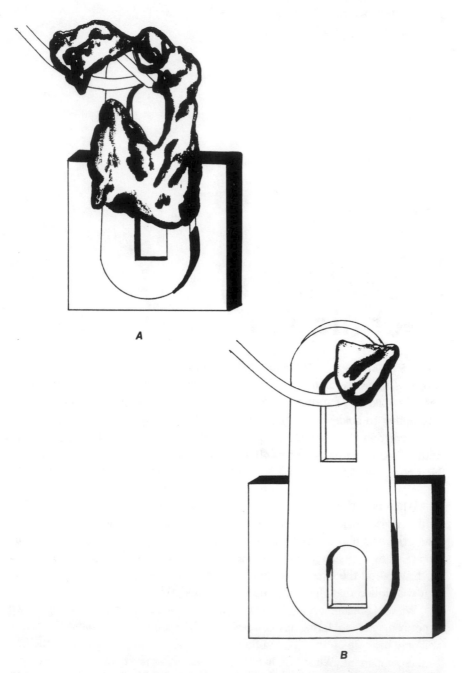

Fig. 5-1. Inspect all soldered connections. (A) shows a cold solder joint that is rough and dull while (B) shows a correctly soldered joint. It is smooth and shiny.

rosin core solder should be used. This type of solder can be purchased from an electronic store or mail-order house.

To ensure a permanent bond between the solder and the wire lead of the component, grease, oil, paint, or any other foreign matter that might be covering the lead must be removed before soldering. To help remove this unwanted material, you can use a steel-wool pad or even a light faced sandpaper.

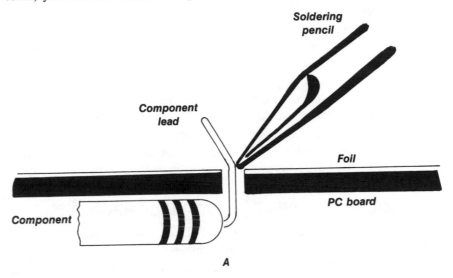

Fig. 5-2A. The correct way to solder a component. Heat the component with soldering iron then . . . .

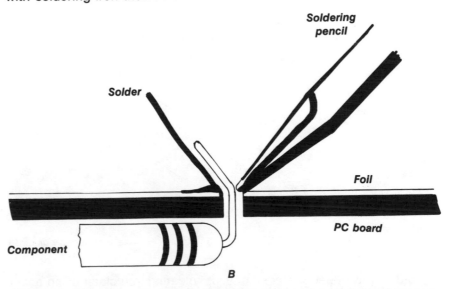

Fig. 5-2B. Have the component and PC board pad melt the solder over the joint.

A major pitfall for the first time solderer is the action of melting the solder with the iron then allowing this molten solder to flow over the component. This is a mortal sin of electronics and should never be done. Using this method will only produce headaches trying to troubleshoot a project that refuses to operate, not because of component failure or improperly designing the PC board, but due to intermittent operation caused by cold solder joints (see Fig. 5-1).

The proper way to solder a component is to bring the tip of the iron in contact with the wire lead and pad of the part, then when heated by the iron, dab a small amount of solder on the component's copper pad. Remove the solder, but leave the iron in place for an additional second or two. This additional heat will allow the solder to flow over the connection point preventing a cold solder joint. The proper soldering procedure can be seen in Fig. 5-2.

After soldering the connection should appear smooth and shiny, while a poor connection will be dull and rough. After soldering a component, the tip of the iron should be cleaned with a damp sponge or cloth to remove excess solder. See Fig. 5-3. You should always keep the soldering iron clean.

## SOLDERING HEAT SENSITIVE COMPONENTS

Integrated Circuits are very sensitive to heat. For this reason, you should solder the leads of an IC fast, to avoid excessive heat that can, and in most cases will destroy the component, not to mention the destruction of the bonding of the copper to the PC board material making the board useless. The use of

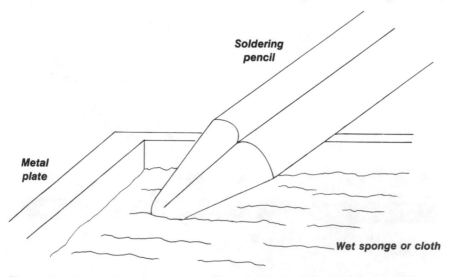

Fig. 5-3. Always keep the point of your soldering iron clean using a wet sponge or cloth.

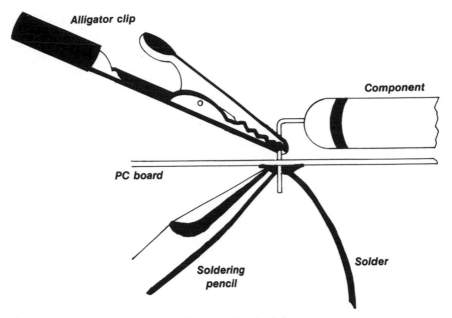

Fig. 5-4. Using an alligator clip as a heat sink.

another type of heat sink is recommended for soldering all heat sensitive components.

The heat sink is a device that is clamped to a component lead to conduct the heat away from the delicate internal circuitry of the part. Heat sinks should also be used when soldering diodes and transistors since these devices are also sensitive to the destructive power of high heat.

Figure 5-4 shows how a common alligator clip can be used as a heat sink. Note how the alligator clip is placed between the component and the soldering iron. By attaching the clip at this location, the heat can be absorbed, preventing the abnormally high temperature created by soldering from being transferred to the component. Needless to say, after using an alligator clip as a heat sink, it will be very hot to the touch and can cause skin burns if you do not exercise caution during its removal.

# 6 Tele-Guard

WITH A THOROUGH DISCUSSION OF TELEPHONE ELECTRONICS IN CHAPTER 1, IN-
telligent component selection, installation, and the fabrication of a printed circuit
board in Chapters 2 and 3, it's time to extend this basic electronic theory into
practical, hands-on experience, by building the first of many telephone
enhancement projects. This will be accomplished by building Tele-Guard, a
homemade telephone bug detector. With such a small number of parts, you can
construct an electronic circuit that has been advertised on television, but can
be constructed with little or no cash layout on the part of the builder.

Tele-Guard detects the presence of an unauthorized listening device on your
telephone line (Tele-Guard also warns the user if an extension telephone is lifted)
and informs the user, by the lighting of a small LED.

## HOW IT WORKS

By referring to Fig. 6-1, the heart of Tele-Guard is an integrated circuit
labeled on the schematic as IC1. From the parts list, IC1 is an LM741, an 8-pin
electronic device. The 741 is a very inexpensive operational amplifier (or op
amp for short) used as a voltage comparator. When a telephone is on-hook,
the dc voltage across the tip and ring (red and green telephone line-cord wires)
is about 50 volts. This voltage can vary depending on the place of your residence
in the country. This 50 volts can be easily seen by connecting a volt meter across

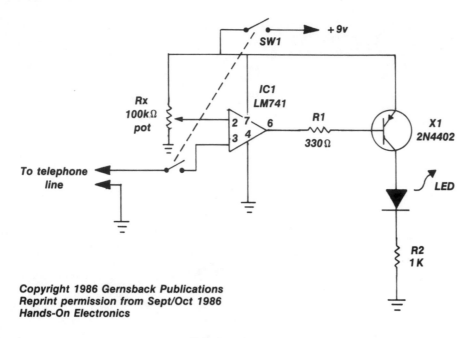

SW1
+ 9v

Rx
100kΩ
pot

IC1
LM741

2 7
3 4
6

R1
330Ω

X1
2N4402

To telephone
line

LED

R2
1K

Copyright 1986 Gernsback Publications
Reprint permission from Sept/Oct 1986
Hands-On Electronics

**Tele-Guard**

Fig. 6-1. Schematic of the telephone bug detector (Tele-Guard).

the line. When the handset is lifted from its cradle, notice that the voltage reading on the meter drops to 5 volts dc. If a telephone listening device is connected across the line or if an extension telephone is lifted, this voltage (5 volts) will drop again even if ever so slightly.

It is this additional voltage drop in the telephone line that Tele-Guard detects, because any bugging device *must* draw voltage from the telephone line for its own operation. With this in mind, to detect the presence of a bug, all you must do is to monitor the normal 5 volt drop and compare this to a reference voltage. This reference is in the form of a 9 volt battery. By referring back to Fig. 6-1, RX is a 100kΩ potentiometer that is user adjustable. This adjustability controls the amount of voltage on pin 2 of the LM741 integrated circuit. With the telephone handset off-hook and no dial-tone heard, this potentiometer is adjusted so that the reference voltage from the battery equals the dc voltage applied to pin 3. The voltage applied to pin 3 is the 5 volts dc across the telephone line when the handset is off-hook.

When the voltage on pins 2 and 3 are equal, the LED is turned on and off by the switching action of transistor X1, that is now off. This indicates that there are no listening devices being used but when there is an imbalance between the same two pins, the LED is turned on by transistor X1. This imbalance is caused by the additional voltage needed by a bugging device or by another telephone when the handset is lifted.

Resistors R1 and R2 are being used as dropping resistors for transistor X1 and the LED. Resistor R2 (1000 ohm) limits the amount of current that the LED can use. If this resistor were not used, the full battery voltage (9 volts) would be applied across the LED. The LED will draw a large amount of current because the internal pn junction has such a small resistance when forward biased. This overflow of current will destroy the junction, thus destroying the LED.

SW1, as seen in the schematic, is a DPDT (double-pole double-throw) switch used to turn Tele-Guard on and off. It also acts as a switch to disconnect the telephone line from IC1. If SW1 is not switched to off, the LED on Tele-Guard will turn on and remain on draining the battery.

IC1 will turn on the LED when the telephone handset is replaced on its cradle, the voltage across the line will increase to its normal 50 volts. With a reference voltage of only 5 volts on pin 2 of the op amp an imbalance is present allowing transistor X1 to conduct and the LED is turned on.

## ASSEMBLY

The assembly of Tele-Guard is not critical. It can be constructed on a solderless breadboard or even a perfboard. But for a more rugged assembly, Fig. 6-3 is a complete printed circuit board layout of Tele-Guard. Figure 6-4 is the component layout, if you wish to use the artwork. This diagram shows the exact placement of all the resistors, transistors, LED, etc., just follow the diagram and insert the proper component indicated. Just remember, the correct mounting and soldering procedure discussed in Chapter 4. If you do, there will be no problems with getting Tele-Guard up and running in no time without problems.

The art layout presented in Fig. 6-3 can be used to construct homemade boards or can be sent to one of many manufacturers that specialize in this type of fabrication. For a more detailed explanation of what is involved in sending out artwork, or making your own PC Board, refer back to Chapters 4 and 5.

*Parts List*

| | |
|---|---|
| Rx | 100K Ohm Potentiometer |
| R1 | 330 Ohm Resistor |
| R2 | 1K Ohm Resistor |
| ICI | LM741 Op-Amp |
| X1 | 2N4402 Transistor |
| LED | Red Light Emitting Diode |
| SW1 | Double Pole Double Throw Switch (DPDT Switch) |

Misc. Parts
Telephone Line Cord
9 Volt Battery

Fig. 6-2. Parts List for Tele-Guard.

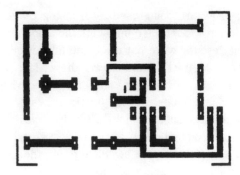

Fig. 6-3. PC Board artwork for Tele-Guard.

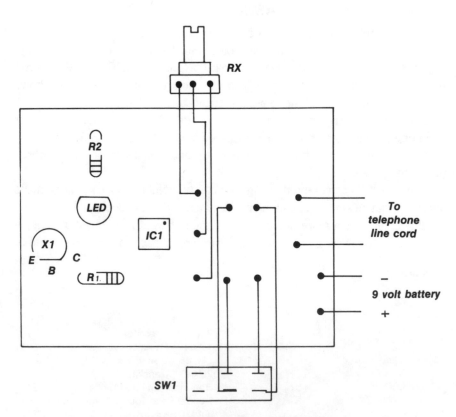

Fig. 6-4. Component layout for Tele-Guard PC board.

Whichever method that is used in construction, be alert to the components that are polarity sensitive. These devices can be installed and soldered *one way* and one way *only*. These components include the transistor (X1), LED, integrated circuit (IC1), and yes, even the telephone line-cord.

## INSTALLATION AND TESTING

With a telephone line-cord connected to the line, as described in Chapter 1, connect Tele-Guard to the line and install the 9-volt battery. Switch SW1 on. Lift the telephone handset. With no dial-tone, adjust RX (100kΩ potentiometer) until the LED lights. From this point, back off a bit until the LED goes out. To test the circuit, lift the handset of an extension. The LED on Tele-Guard will light up indicating someone is listening to your conversation. Hang up the extension, the LED will now go off. If the LED does not light, reverse the telephone line-cord wires (red/green) on Tele-Guard. Remember, when not in use, turn SW1 to the off position.

# 7 Telephone Hold Button

THE IDEA OF PLACING A TELEPHONE CALL ON HOLD IS NOT NEW. LARGE INSTITU-
tions have been doing it for years. Recently, circuits have been designed (KSU—
key service units) that allow the user to connect AM/FM radios, tape decks
or any other audio signal to the equipment, when a party is placed on hold, the
caller is entertained by a musical interlude rather than listening to silence.

Today, electronically produced music has done away with the cumbersome
radio and cables needed for *music-on-hold* (MOH). To generate the music needed
by the KSU's, a small microcomputer chip will be used, and best of all, this
chip is readily available in a greeting card store, not an electronic mail order
house. I am describing a musical greeting card. Go to a card shop and open
one of these electronic marvels. You will hear the card play a musical tune for
about 30 seconds then repeat. A closer inspection will reveal the microcomputer
circuit that operates from a 1.5 volt battery with a miniature transducer (another name for a speaker) wired in. When the card is opened a small plastic band
moves from between a copper PC board pad and a metal strip, allowing the
strip to make contact, is a basic on/off switch. With the switch closed, power
from the 1.5 volt battery is supplied to the circuit causing the computer to play
the pre-programmed melody.

After purchasing a musical greeting card, carefully remove the microcomputer chip and transducer from the cardboard mounting and place them aside

**68**

for now. For this synthesizer to operate with the telephone hold button slight modifications must be made, but more on that a little later.

## HOW IT WORKS

The voltage across a standard telephone line (red/green wires) is 50 volts dc. When the handset is lifted from its cradle, the impedence of the telephone (600 ohms) is placed across the line through the closing action of the internal telephone hookswitch. This resistance drops the line voltage from 50 volts to 5 volts dc. To place a call on hold, all you must do is to fool the central office switching equipment (network) into thinking that a telephone is still on line even if it's not. To do this, a resistive load is placed across the line. This load is in the form of a 100 to 1000 ohm resistor.

To give the user a visual indication that a call is on hold, some kind of external indicator is needed. To keep the coat of the telephone hold button to a minimum, an inexpensive LED will be used. The circuit presented in Fig. 7-1 is a schematic

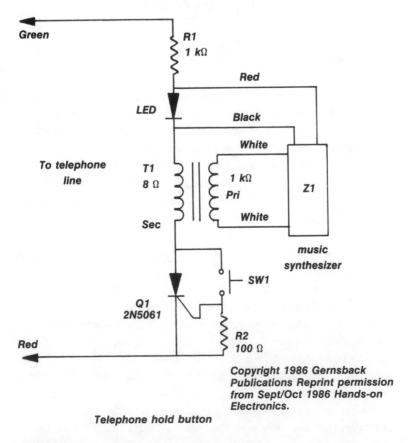

Copyright 1986 Gernsback Publications Reprint permission from Sept/Oct 1986 Hands-on Electronics.

Fig. 7-1. Schematic of the Telephone Hold Button.

***Parts List***

| | |
|---|---|
| R1 | 1K Ohm Resistor |
| R2 | 100 Ohm Resistor |
| LED | Red Light Emitting Diode |
| T1 | Audio Transformer<br>1.2K Ohms to 8 Ohms |
| Q1 | 2N5061 or Equiv. SCR |
| Z1 | Modified Music Synthesizer |
| SW1 | Single Pole Single Throw<br>(SPST) Push Button Switch |
| Misc. Parts | |
| Telephone Line Cord | |

Fig. 7-2. Parts list for the Telephone Hold Button.

diagram of the telephone hold button. The part list is in Fig. 7-2. The resistive load mentioned earlier that will keep a party on hold is in the resistor R1 (1000 ohm resistor). The "on hold" indicator is the series LED.

To transfer the output of the music synthesizer across the telephone line and also to isolate it from the line voltage, a transformer (T1) is used. The transformer is a standard audio type with a primary impedence of 1200 ohms and a secondary impedence of 8 ohms. The audio output of the music synthesizer is connected to the primary of the transformer, while the secondary is connected in series with R1, the LED, and the Q1 transistor.

The component of Q1 is a SCR. This transistor will be used as a solid state switch. When conducting, Q1 will allow the telephone line current to flow between the red and green line-cord wires, then with the music synthesizer playing its tune, this melody is impressed in the telephone line as well as the resistance of R1.

To turn on the SCR (Q1), switch SW1 (a spring-return-type push-button) is closed. The voltage from the telephone line is allowed to enter the gate of Q1 causing the SCR to conduct, this completes the series circuit. Now a resistive load of 1000 ohms is electronically connected across the telephone line. While still pressing SW1, the user will now be able to hang up the telephone without disconnecting the call.

To reconnect the call, all you must do is pick up the handset again. This action will stop the SCR from conducting by disconnecting R1 and the LED from the line. The music will stop and the LED will go out.

The music synthesizer (Z1 in Fig. 7-1) requires 1.5 volts for normal operation. With a call on hold, the LED is lit. As an experiment, place a voltmeter across the two wire leads of the LED and note the voltage. It will be about 1.5 volts. What if two wires from the module are soldered across the LED leads? The voltage drop across the LED, and the low current required by the music synthesizer make this wiring ideal. The module is powered by your local telephone company.

## MUSIC SYNTHESIZER MODIFICATIONS

To make use of the synthesizer with the telephone hold button, slight modifications must be made to the board. For now, refer to Fig. 7-3. This drawing represents a typical music synthesizer that can be found inside an imported greeting card. Note that the styling may differ from card to card but the modifications are still the same. In the upper left-hand corner of Fig. 7-3 is the basic on/off switch, this must be soldered closed. If not, after a while oxidation will form on the copper trace causing intermittent operation.

In the center right-hand side, the 1.5 volt battery is located, because the voltage needed for operation is received from the project board, this battery can be removed and two wires soldered in its place. The top portion of the battery holder is for positive voltage, so carefully solder a *red* wire to this location. The section directly beneath is for negative voltage. So at this location, solder a *black* wire. While soldering, be careful not to overheat the module. After all, this is a microchip, excess heat can spell destruction for the delicate circuitry.

The audio transducer that is wired to the microchip can remain connected or be removed depending on the project being constructed. The following project Telephone Melody Ringer also makes use of the microchip. The transducer

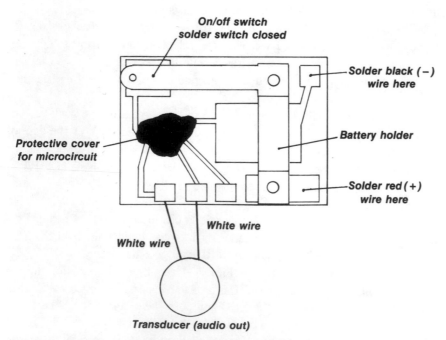

Fig. 7-3. A typical greeting card music generator IC.

for this project can remain connected, but for the telephone hold button it must be removed, cut the transducer off at the transducer end of the wires. These *white* wires from the synthesizer's audio output will be used to connect the chip to the hold button board.

## ASSEMBLY

The assembly of the hold button is not critical. You can make use of a perforated board or even a solderless experimenter board. But for those who wish to make their own printed circuit boards, Fig. 7-4 is the full size PC board artwork, and in Fig. 7-5 is the parts placement diagram for the telephone hold button. If you wish to make your own boards, please refer to Chapter 4 for additional information. Whichever means is chosen, remember to keep in mind the proper assembly procedures described in Chapter 2. Keep all leads short, use heat sinks when soldering the transistor and the LED. And pay strict attention to the proper insertion of components.

## INSTALLATION AND TESTING

With the telephone hold button connected to the dial-tone line (connect the hold button in parallel with the red and green wires located inside the telephone wall jack), lift the handset from the cradle and press SW1. The tune generated by the synthesizer should now be heard, and the LED is also lit, but not to full brilliance. If the circuit seems dead, simply reverse the red and green wires of the telephone line-cord that is connected to the hold button.

*Copyright 1986, Gernsback Publications, reprint permission Sept/Oct 1986, Hands-on Electronics.*

*Telephone hold button w/music synthesizer*

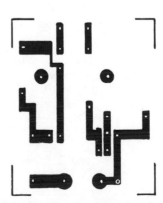

Fig. 7-4. PC board artwork for the Telephone Hold Button.

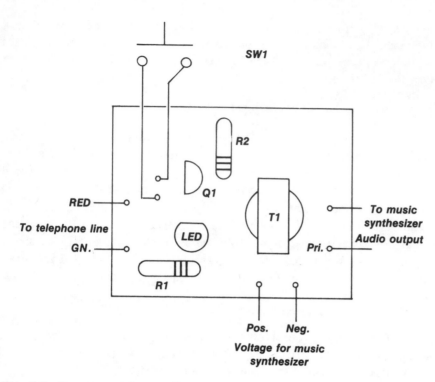

Fig. 7-5. Component layout for the Telephone Hold Button PC board.

To place a call on hold, just press SW1. While still pressing SW1, return the handset to its cradle. The LED will now glow brighter and stay on. To reconnect the holding party, simply lift the handset from the same telephone or other extension on the same line. The LED will go off and the music will stop.

If the LED remains lit even after lifting the handset from the receiver, the value of R2 must be changed. This is due to the fact that telephone companies across the country provide different line voltage to their phones. In New York, this line voltage can be 50 volts (on hook) and in Florida, the same line voltage could drop to 35 walts (also on hook). For this reason, you might experience a problem with the hold button. To rectify this situation, simply adjust the value of R2 until the LED extinguishes when a party on hold is to be reconnected.

# 8 Telephone Melody Ringer

TECHNOLOGY IS CHANGING RAPIDLY, BELLS USED IN TELEPHONES SEEM TO BE A thing of the past. Specifically designed integrated circuits have replaced the annoying mechanical device with the soft chirp of state of the art electronics.

The next project presented converts a standard bell-type telephone into a customized melody ringer. There's no need to purchase an imported telephone add on device, the telephone melody ringer can be constructed in about one half-hour and it might not even cost a dime to construct. All the components needed for construction could already be in your parts drawer. To cut the cost of the ringer even further, the music synthesizer used in the last project (Telephone Hold Button), can be put to use here also. The cost of the melody greeting card is so inexpensive, why not just purchase two. Use one synthesizer for the hold button and the second for the melody ringer.

Whichever route is decided upon, the Music Synthesizer must be modified as discussed in Chapter 7. So go back and refresh your memory before attempting to make the changes to the delicate microchip. The 1.5 volt battery that comes with the greeting card is not used with this project and can be removed. The voltage necessary to power the microchip is provided by another 9 volt battery.

## HOW IT WORKS

For the following, refer to the schematic diagram presented in Fig. 8-1. Connect the finished melody ringer in parallel to the telephone line (red/green

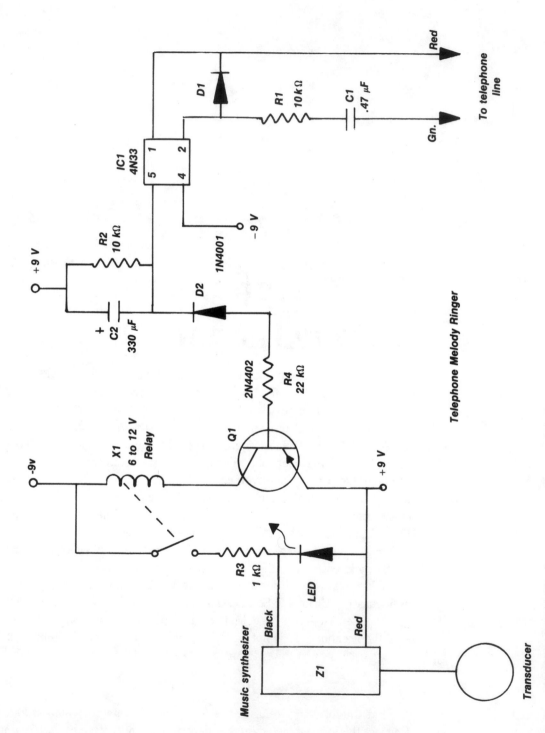

Fig. 8-1. Schematic of the Telephone Melody Ringer.

wires) as shown. When a call is being received, there is a 120 volt ac 30 Hz signal across these two wires. This large voltage is applied to the internal bell that rings in sequence with the 20 or 30 Hz sine wave. To detect this large ringing voltage, an opto-isolator IC (4N33) is used. This device incorporates a small LED that flashes on a light sensitive Darlington transistor circuit. When a voltage is applied to pins 1 and 2 (LED input) the output on pins 4 and 5 go low. Lets examine what happens when a telephone call is received.

The 120 volt ac ringing signal is applied to the circuit via the telephone line-cord. During its travels, the ringing signal meets up with C1. A 0.47 $\mu$F, 250 volt capacitor. This capacitor is used to block the normal telephone dc voltage from coming into the IC (4N33) keeping its internal LED always on. If this happens, the music synthesizer will constantly play even if there is no ringing voltage present.

A standard LED draws a small amount of current, a 10k$\Omega$ resistor (R1) can be placed in series with the telephone line to knock the large ringing voltage down to acceptable voltage levels that can be handled easily by the LED that is located inside the IC1. This resistor can be the ¼ watt carbon type.

The diode referred to as D1, conducts every half-cycle of the incoming ring signal. If an oscilloscope is placed across pins 1 and 2 of the opto-isolator IC, you will see a pulsating dc voltage. This signal resembles a half-wave rectifier. This waveform is discussed in Chapter 2. If you have skipped over this chapter, you could find it helpful to go back for a brief review.

*Parts List*

| | |
|---|---|
| R1 R2 | 10K Ohm Resistor |
| R3 | 1K Ohm Resistor |
| R4 | 22K Ohm Resistor |
| C1 | .47uF Capacitor |
| C2 | 330uF 16V Capacitor |
| D1 D2 | 1N4001 Diode |
| Q1 | 2N4402 Transistor |
| IC1 | 4N33 Opto-Isolator |
| X1 | 6 to 12 volt relay |
| | Coil resistance not below |
| | 700 Ohms |
| LED | Red Light Emitting Diode |
| Misc. Parts | |
| Z1 | Modified Music Synthesizer |
| 1 | Crystal Transducer (Comes |
| | soldered on Synthesizer) |
| 1 | 9Volt Battery |
| 1 | Telephone Line Cord |

Fig. 8-2. Parts Lists for the Telephone Melody Ringer.

As indicated earlier, with a signal applied to pins 1 and 2 (IC1) the output (pins 4 and 5) goes low. The negative side of the 9 volt battery can now be allowed to pass through the Darlington transistors inside the 4N33 IC and charge up capacitor C2.

During the "off" cycle of the input ring, this capacitor discharges through R2 and D2. The 10kΩ resistor (R2) allows the C2 capacitor to discharge slowly. The discharging rate is about 4 seconds. It is this 4 second delay that prevents the X1 relay from dropping out as soon as the telephone ringing signal is in its off state, this is also about 4 seconds.

The voltage of the discharging capacitor is applied to the base of transistor Q1 (2N4402) turning it on. The collector of Q1 is connected to a SPST (signal pole single throw) relay (X1) when activated by the now conducting transistor (Q1), closes its contacts.

The contacts of the X1 relay is connected in such a way as to deliver 9 volts to an external LED through a dropping resistor (R3). The power needed for the operation of the music synthesizer is obtained like the telephone hold button, across the LED. With this configuration, every time the telephone rings, the X1 relay pulls in, applying power to the music synthesizer so a tune can be heard.

The only thing to take into consideration is that the resistance of the relay coil be no less than 700 ohms. If a lower resistance is used, the discharge delay of 4 seconds will no longer hold true. With a lower resistance, capacitor C2 will discharge faster allowing the X1 relay to drop out during the off-cycle of the ringing signal.

With a telephone now ringing, you'll find that the volume of the music generator is too low, you can substitute a crystal telephone receiver element for the audio transducer that comes with the synthesizer. This substitution will provide a louder output without the added expense of an amplifier. Just remember to remove the varistor component that is soldered across the two screw terminals of the receiver.

## ASSEMBLY

Like the other projects in this book, the assembly of the melody ringer is not critical. You can wire the ringer on a perf board or solderless experimenter board. For those who wish to make their own PC boards, Fig. 8-3 illustrates the circuit connections that can be photographed or traced to provide the final artwork needed for the finished board. Figure 8-4 provides the component layout for the artwork. Note that the X1 relay is not mounted on the PC board. There are a large number of relays that can be used in this circuit, I have chosen not to prevent the hobbyist from using available relays by designing the artwork for a particular type or make of relay. So make use of what is available to keep the overall cost to a minimum.

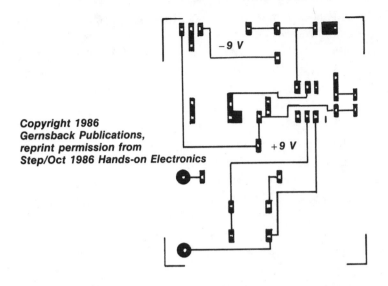

Copyright 1986
Gernsback Publications,
reprint permission from
Step/Oct 1986 Hands-on Electronics

*Melody Ringer*

Fig. 8-3. PC board artwork for the Melody Ringer.

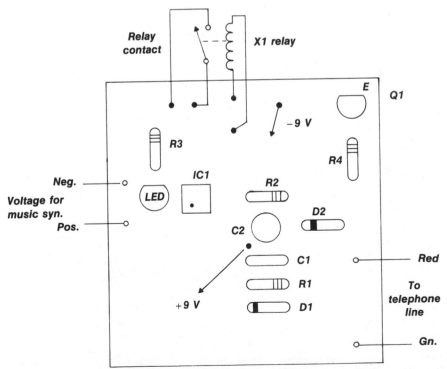

Fig. 8-4. Component layout for the Melody Ringer PC board.

Pay close attention to the polarity of C2, both diodes (D1 and D2), transistor Q1, the LED, IC and of course the Music Synthesizer. If any problems develop in the operation of the Melody Ringer, you should first look for improperly inserted components and solder bridges. If the problem is still not found, put the board aside for a while. Return to the project in an hour or two. You can be very surprised at the silly errors you might have overlooked previously.

## INSTALLATION AND TESTING

With a 9 volt battery connected to the Melody Ringer and the ringer is connected to the telephone line, short pins 4 and 5 of the opto-isolator (IC1) together. This short will provide a path from the negative side of the battery to the parallel combination of C2 and R2 charging them. At this time, the X1 relay will pull in and a musical interlude will be heard from the transducer. Remove the short, notice that the tune will continue to play for an additional 3 to 4 seconds. This is the discharge delay I discussed earlier. This is a normal function of the circuit.

After the above test, the Melody Ringer can now be permanently connected to the telephone line. Then when a call is received, you can be serenaded with the song of your choice.

# 9 Telephone Tone Ringer

UNLIKE THE MORE ELABORATE MELODY RINGER (SEE CHAPTER 8) THE TELE-
phone Tone Ringer presented here will not play today's hit music. Instead it
will convert the harsh ringing of a bell into a more soothing tone output.

## HOW IT WORKS

The schematic presented in Fig. 9-1 is the basic wiring configuration for
the Motorola MC34012 tone generator. There are three available ICs each having
a different base frequency. They are as follows:

| | |
|---|---|
| MC34012-1 | 1.0 kHz. |
| MC34012-2 | 2.0 kHz. |
| MC34012-3 | 500 Hz. |

Any of the three tone ICs receives its operating voltage from rectifying the
ac ring signal that has its input at pins 2 and 3. The IC uses this voltage to activate
an internal tone generator which drives an outboard piezo-ceramic transducer
or a telephone receiver element. The tone generator circuitry includes a
relaxation oscillator and frequency dividers that produce high and low frequency
tones as well as the tone warble frequency. The frequency of the main oscillator

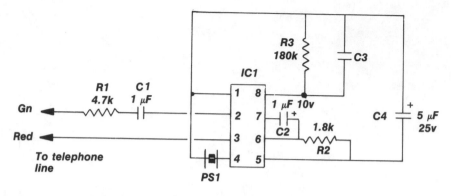

Fig. 9-1. Schematic of the Telephone Tone Ringer.

**Parts List**

| | |
|---|---|
| R1 | 4.7K Ohm Resistor |
| R2 | 1.8K Ohm Resistor |
| R3 | 180K Ohm Resistor |
| C1 | 1μf Capacitor (Nonpolar) |
| C2 | 1μf 10V Capacitor |
| C3 | See Text |
| IC1 | MC34012-1 or |
| | MC34012-2 or |
| | MC34012-3 |
| PS1 | Piezo Sound Element or |
| | Telephone receiver |
| | element |
| Misc. Parts | |
| 1 | Telephone Line Cord |

Fig. 9-2. Parts list for the Tone Ringer.

is set by the values of R3 and C3. The value of C3 can be the following, keep in mind the type of IC purchased.

| | |
|---|---|
| MC34012-1 | C3 = 1000 pF |
| MC34012-2 | C3 = 500 pF |
| MC34012-3 | C3 = 2000 pF |

The values of C3 are only recommendations, experimentation takes place with the value of C3 as well as the value of R3 (180 kΩ) to obtain the desired output tone. Input signal detection circuitry activates the tone ringer output when the ac ring voltage exceeds programmed threshold levels. The value of resistor R2 determines this level. The value of R2 can be in the range of 800 to 2.0 kΩ.

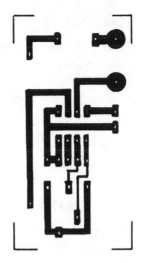

*Telephone tone ringer*

Fig. 9-3. PC board artwork for the Tone Ringer Project.

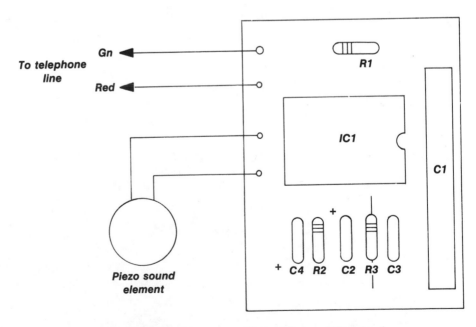

Fig. 9-4. Component layout for the Tone Ringer PC board.

## ASSEMBLY

The assembly of the telephone tone ringer will prove to be quite simple if the PC board artwork presented in Fig. 9-3 and the component layout diagram in Fig. 9-4 is used. Just be careful of component polarities, and when soldering *do not* overheat the IC. Use an alligator clip as a heat sink. This will pull the destructive heat generated by soldering, away from the delicate integrated circuit.

The tone ringer is simple enough that a PC board can be eliminated so that time, and the cost of etching material can be saved. A perforated board can also be used.

## INSTALLATION

Because the tone ringer is so small, you may want to consider mounting it directly inside a telephone. To do this, remove the housing. You might consider removing the bell from the phone by first locating the two or four wires that are connected to the network. Remove these wires and unscrew the bell.

Using double-faced tape (sticky on both sides), fasten the tone ringer to the base of the telephone. After mounting, connect the two dial-tone wires from the tone ringer to the green and red telephone-line input wires (these wires can be located on the L1 or L2 terminal on the network). Replace the telephone housing and ask a friend to ring your phone to test the circuit. You will be surprised how pleasant a telephone could sound with the right enhancement project installed when ringing.

# 10    Automatic Telephone Recorder

CHAPTER 6 PRESENTED A PROJECT THAT DETECTED BUGS ON A TELEPHONE LINE. The Automatic Telephone Recorder of this chapter will give Tele-Guard something to detect. The automatic telephone recorder is a device installed in the main telephone line and its purpose is to record telephone conversations from any telephone connected to a common line (telephone extension using the same telephone number).

It will automatically start a tape machine and record both sides of the conversation when the handset of any telephone on-line is lifted. Note that the cassette recorder must be left in the *record* mode for proper operation.

## HOW IT WORKS

For the following, refer to Fig. 10-1 the schematic diagram of the automatic telephone recorder. The recorder does not need any battery for operation. The required voltage comes from the telephone line itself. It might not be easy to see, but the telephone recorder is basically a small relay wired in series with the incoming telephone line. When all house telephones are on-hook, there is 50 volts applied across the red and green wires of the line-cord. The X1 relay will not energize because the internal contacts of the telephone (also known as the hook-switch) are in the open position, preventing the resistance of the phone from being connected to the line. Seeing that this is an open circuit, the

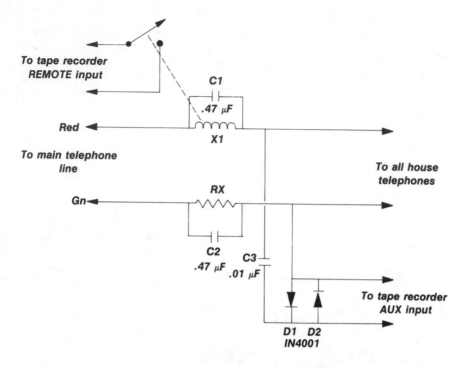

To tape recorder
REMOTE input

C1
.47 µF

Red

X1

To main telephone
line

To all house
telephones

RX

Gn

C2
.47 µF

C3
.01 µF

D1  D2
IN4001

To tape recorder
AUX input

Fig. 10-1. Schematic of the Automatic Telephone Recorder.

negative 50 volts in the red line-cord wire cannot complete the journey to the positive or green line-cord wire because of the open hook-switch contacts.

When a telephone handset is lifted, the hook-switch is closed completing the series circuit. The once 50 volts across the line, now drops to about 5 volts. Because the X1 relay is connected within this now closed electrical loop, it energizes closing the contacts.

The contacts of the relay are wired and brought out to a connector jack, where it is plugged into the remote input of the cassette recorder. The X1 relay contacts now control the motor of the recorder. When the contacts close, the motor starts and when the contacts are open, the motor stops. This on-and-off action is controlled by the lifting and replacing of the telephone handset.

For the proper operation of the telephone recorder, the X1 relay must be a low voltage type. A DIP relay with a voltage rating of 5 volts and a coil resistance of about 100 ohms. To record the conversation using the automatic telephone recorder, all that is needed is capacitor C3. This .01 µF capacitor, blocks the 5 volts dc from entering the cassette recorders *Aux* input, but allows the audio signal. It is this signal that is recorded.

When the telephone conversation is complete and the handset is returned to the cradle, the relay (X1) loses its voltage and drops out. This opens the

*Parts List*

| | |
|---|---|
| C1 C2 | .47µf 250 volt Capacitor |
| C3 | .01µf Capacitor |
| X1 | See Text |
| RX | See Text |
| D1 D2 | 1N4001 Diodes |
| Miss. Parts | |
| Connector Jacks to match | |
| tape recorder. | |
| Telephone Line Cord | |
| Housing | |

Fig. 10-2. Parts list for the Telephone Recorder.

relay contacts and the recorder stops and waits for the next call. The capacitors (C1 and C2) and the diodes (D1 and D2) are used to route the ringing voltage around sensitive electronics that could be destroyed by this 120 volt ac signal.

The ring signal is a high ac voltage. By using C1 and C2, this signal is routed around the coil of the relay and resistor RX. Any electronic signal, will follow a path of the least amount of resistance. The coil of X1 has a resistance. RX has resistance, so the ring voltage uses the easy path by flowing through C1 and C2 instead.

The 1N4001 diodes (D1 and D2) have the same function as the capacitors. They prevent the ac signal from entering the cassette recorders *aux* input.

## MAINTAINING A BALANCED LINE

Whenever any outside equipment is connected to a telephone line, the line must maintain a balanced voltage. A relay coil is being connected to the ring (red wire) side of the line, there must be an equal resistance on the *tip* (green wire) side. And this is the function of resistor RX. To determine the value of RX, measure the resistance of the X1 relay coil. This value can be about 100 ohms. The value of RX must be 100 ohms. If the resistance is higher, the value of RX must match it.

## ASSEMBLY

The assembly of the Telephone Recorder is not critical. But it is recommended that the finished project be contained in a plastic housing. This is to protect the electronics, because of the location that the recorder is to be installed. Even before the assembly begins, you should determine what cassette recorder is to be used. When this is decided, mating connectors for the recorders *REMOTE* and *AUX* input should be purchased. The assembly of the telephone recorder is straightforward and should cause no problems.

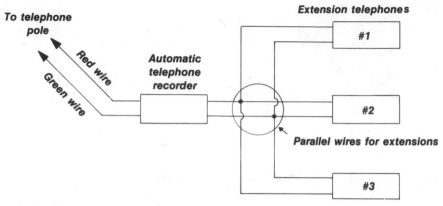

Fig. 10-3. Typical installation of the Telephone Recorder.

## INSTALLATION

When the assembly is complete and mounted in a plastic box, determine a location where the recorder can be installed. This location must be in the main telephone line before any parallel connections were made for extensions, and where the cassette recorder is easily accessible.

When the location is found, cut the red telephone line wire and connect this half to the red wire input of the recorder (to the main telephone line). The second half of the red wire is to be connected to the red output of the recorder (to all house telephones). When complete, do the same for the green wire of the main telephone line (see Fig. 10-3). Connect the telephone recorder in *series* with the existing telephone line.

When complete, connect the jacks to the appropriate cassette recorder input. Place the tape machine in the record mode. To test the circuit, lift the handset of any telephone and notice that the tape cassette is turning in the machine. Return the handset to the cradle, the cassette recorder should now be off. To listen to a recorded tape, just rewind the cassette and play in the usual manner.

# 11   Telephone Call Indicator

NEVER LEAVE YOUR HOUSE OR APARTMENT THEN RETURN WONDERING IF THAT important telephone call ever came? With the Telephone Call Indicator, wonder no more.

By incorporating this simple circuit in the telephone line, a small LED will light, indicating a call was received in your absence. The LED will remain on until reset by pressing SW1. The circuit may not be a sophisticated telephone answering machine but many will find a usefulness for the project. Whether the circuit detects a ringing telephone or is used as a building block for a project of your own design, the telephone call indicator is well worth the time invested in its assembly and installation.

Figure 11-1 is the complete schematic diagram for the detector. It is relatively simple and straightforward. The heart of the circuit is the SCR (Silicon Controlled Rectifier). When a small amount of voltage is delivered to its gate (the lead designation for an SCR is gate, anode and cathode) the Anode and Cathode leads allow the voltage from the 9 volt battery to flow, thus lighting the LED. The most significant feature is that an SCR continues to allow a current to flow even if the applied voltage of the gate is removed. In this project, the LED indicates this current flow. The LED will remain on until the reset button (SW1) is pressed disconnecting the voltage from the SCR. When the voltage is removed, the LED will extinguish and remain out until another telephone call is detected.

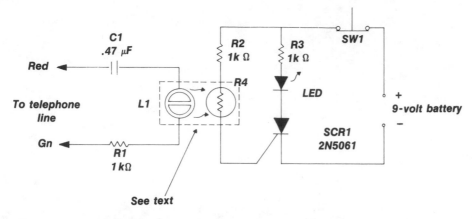

Fig. 11-1. Schematic of the Telephone Call Indicator.

## RING DETECTION

For the circuit to function as a call indicator, it must know when a call is being received. This is the job for L1 and R4, L1 is a neon light. Unlike standard 117 Vac light bulbs that radiate light even with a small amount of voltage being applied, the neon will only display visible light when its firing point is reached. This firing point is about 90 volts.

If a voltage of 50 volts were to be supplied, the neon bulb will not show any indication of light. Light will only be produced when a voltage of 90 or greater is applied. Like an LED, the neon bulb must have a dropping resistor placed in series to protect it from high voltage levels. In Fig. 11-1, this dropping resistor is in the form of R1 (1000 ohm).

A ringing signal of a telephone is 120 volts ac, the neon bulb is the perfect way to detect an incoming call. So if a series combination of R1 (1000 ohm resistor) L1 (neon bulb) and an isolation capacitor (C1 - .47 $\mu$F 250 volt) is placed in parallel across the telephone line, the neon bulb will flash every time the telephone line voltage exceeds 90 volts. This will happen every time a call is being received. Now with L1 flashing, a means had to be found to somehow interconnect the flashing neon to the gate of SCR1. For this, R4, a *photocell* is used.

A photocell is really a variable resistor, unlike a potentiometer, a photocell's resistance is dependent on the amount of visible light that is placed on its facing material. With no light shining on the face, some photocells have a resistance in the millions of ohms, one can also say that in darkness, the photocell is an open circuit. When light is applied, the cell's resistance drops proportionally to the amount of light. If a battery is connected to a cell through a resistor, in darkness, there will be 0 voltage. If a light is applied to the face of the cell, the internal resistance will drop allowing a current to flow.

*Parts List*

| | |
|---|---|
| R1   R2   R3 | 1000 Ohm Resistors |
| R4 | Photocell |
| C1 | .47$\mu$f 250 volt Capacitor |
| SCR1 | 2N5061 transistor or equiv. |
| LED | LED (Light Emitting Diode) |
| SW1 | Spring return push button switch with normally closed contacts |
| L1 | Neon bulb |
| Misc. Parts | |
| 1 | 9 volt battery |
| 1 | telephone line cord |
| Black Electrical Tape | |

Fig. 11-2. Parts list for the Call Indicator.

## MAKING A RING DETECTOR

With this information, you can easily make a telephone ring detector by taping (use black electrical tape) the neon bulb (L1) to the face of the photocell (R4), making sure this combination is *light tight*. This assembly can be seen in Fig. 11-3. Remember to allow the leads of the components to protrude from the tape.

## HOW IT WORKS

With a ringing signal (120 Vac at 30 Hz) applied to the red and green wires of the call indicator, L1 (neon bulb) starts to flash. This flashing light reduces the resistance of the photocell (R4). With a lower resistance, a current is allowed to flow to the gate of SCR1. The electron flow on the gate turns the SCR on, thus lighting the LED. This LED is the indicator that shows a telephone call was received. The LED will remain on until it is extinguished by pressing the normally closed push button switch (SW1).

## ASSEMBLY

This project will be sitting next to a telephone, you will want the call indicator mounted in any eye pleasing housing. The type of housing will depend on the individual using the project. The call indicator is an unsophisticated project that requires very little cash layout on the part of the hobbyist.

Just remember to pay attention to the installation of the SCR and LED. These components can be damaged if inserted improperly. In Fig. 11-4 is a PC board layout of the telephone call indicator that can be used to make your own boards. Figure 11-5 is the physical component layout for the artwork. This is the perfect starter project for the hobbyist wanting a useful circuit that can be assembled in no time at all.

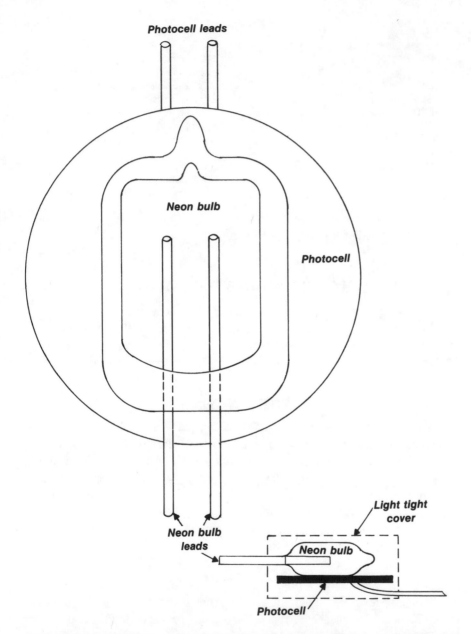

Fig. 11-3. How to make a Ring Detector using a neon bulb and a photo-cell.

## INSTALLATION

With the Call Indicator assembled and mounted in the housing of your choice, connect the RED and GN project line-cord to the corresponding red and green wires in the telephone junction box, or if you have a junction box using the new module connector, just plug the project line-cord into the box.

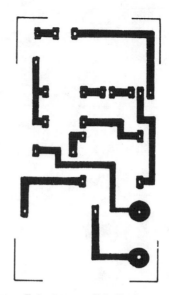

*Telephone call indicator*

Fig. 11-4. PC board artwork for the Telephone Call Indicator.

To test the circuit without waiting for a telephone call, take a length of wire and short out the photocell (R4). This will allow a current to flow to the gate of the SCR lighting the LED. Remove the short and the LED will remain on. To extinguish, press SW1.

When a call is received, recheck to see if the LED is on, if not, recheck the wiring for components C1, L1 and R1. If all seems fine, check to see if the light from the flashing neon light is falling on the photocell. If you wish, an on/off switch can be installed in either the positive or negative voltage supply.

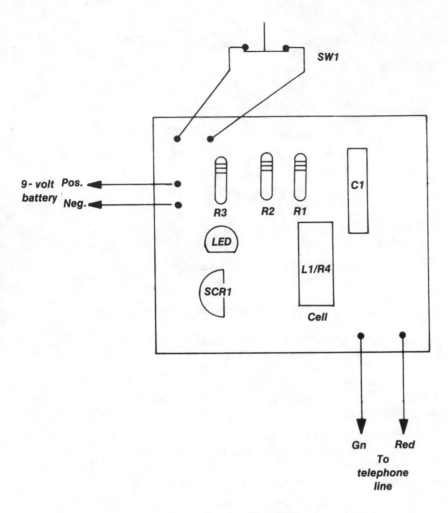

Fig. 11-5. Component Layout for the Call Indicator project.

# 12 Telephone Ring Indicators

PRESENTED WITHIN THIS CHAPTER ARE TWO USEFUL BUT ULTRA-SIMPLE VISUAL telephone ring indicators. Both circuits convert the annoying harsh sounding bell into the quiet flashing of a lamp. The projects are a perfect addition to the nursery where loud noises can wake a sleeping baby.

## HOW IT WORKS (LED INDICATOR)

Figure 12-1 shows a schematic of a ring indicator using an LED that flashes when a telephone call is received. With this 120 Vac 30 Hz sine wave being applied, D1 (1N914 Diode) converts the ac into a pulsating dc voltage. By referring to Chapter 2, a detailed explanation on diode rectification is presented. By reviewing this chapter you can easily understand the conversion of an ac signal into a pulsating dc voltage.

Resistor R1 serves as a dropping resistor to protect the LED from high current that can destroy the delicate internal pn junction. With the telephone ring signal limited in current by R1 and rectified by D1, this voltage is safely applied to the LED where the LED will flash whenever a call is received.

## ASSEMBLY

The assembly of the LED call indicator will take a hobbyist about 5 minutes to put together. Just remember to observe the polarity of the diode and the

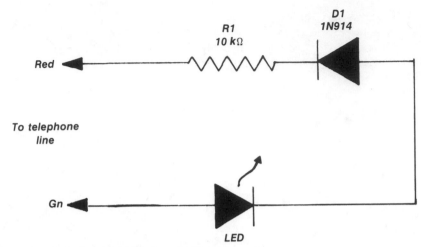

Fig. 12-1. A simple LED ring indicator.

**Parts List**

| R1 | 10K Ohm Resistor |
|---|---|
| D1 | 1N914 Diode or Equiv. |
| LED | Light Emitting Diode |
| Misc. Parts | |
| 1 | Telephone Line Cord |
| 1 | Optional Housing |

Fig. 12-2. Parts list for the LED Ring Indicator.

LED. You can assemble the indicators on a perforated board or use the artwork provided in Fig. 12-5 to etch a PC board.

## HOW IT WORKS (NEON LAMP INDICATOR)

Figure 12-3 is another simple ring indicator fashioned from a neon lamp. As with the LED indicator (Fig. 12-1), a dropping resistor must be included in the circuit to limit the amount of current drawn by L1.

The D1 diode presented in Fig. 12-1 is mysteriously missing from Fig. 12-3. It's really no mystery. The neon lamp (L1) will operate better and provide additional light using an ac voltage instead of dc. This difference can be seen when a telephone call is being received. Look closely at the internal electrodes of the neon lamp. Notice that both electrodes glow with an ac voltage applied. Now, place a diode in series with L1 and observe that only one of the electrodes is glowing. This is due to the changing polarities of the electrodes with an ac voltage. But if dc is applied, one electrode will remain positive while the second will remain negative. Unlike the LED indicator that provides a small amount

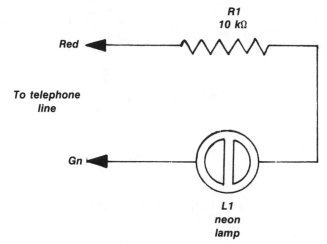

Fig. 12-3. A simple neon bulb can be a great Ring Indicator.

**Parts List**

| | |
|---|---|
| R1 | 10K Ohm Resistor |
| L1 | Neon Lamp |
| Misc. Parts | |
| 1 | Telephone Line Cord |
| 1 | Optional Housing |

Fig. 12-4. Parts list for the neon bulb Ring Indicator.

of light when energized by a ring signal, the neon lamp will glow brighter and provide a more attention getting light.

## PRINTED CIRCUIT BOARD ASSEMBLY

Figure 12-5 provides a simple layout for the etching of a printed circuit board for the telephone ring indicators. The artwork can be used for the LED indicator. The board can also be used for the neon indicator if the space provided for diode D1 is shorted using a piece of wire.

## INSTALLATION

To install the indicators, all you must do is connect the circuit line-cord to the corresponding red and green wires located in the telephone junction box mounted on the wall near the floor. When complete, the ringing of a telephone will be replaced by the soft glow of light.

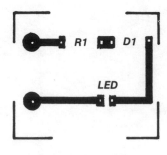

**LED Ring Indicator**

Fig. 12-5. PC board artwork that can be used for both LED and neon bulb Ring Indicators.

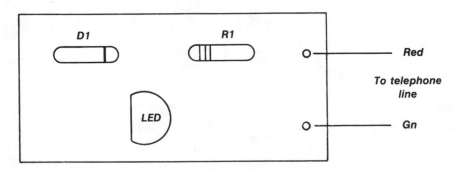

Fig. 12-6A. Component layout for the LED Ring Indicator.

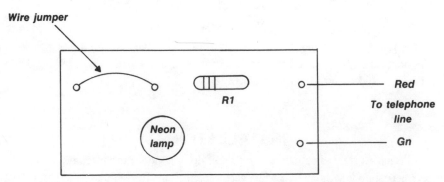

Fig. 12-6B. Component layout for the Neon Bulb Indicator.

## Another Installation Idea

Because the indicators are so small you might consider mounting them directly inside the telephone. Just drill a hole in the housing near the top and install either the LED or neon lamp.

To disable the telephone ringer, just trace the 2 or 4 wires coming from the bell and disconnect one wire from its location on the network. When disconnected, tape the lead to prevent shorting to any other part inside the housing.

# 13   Conference Caller

A CONFERENCE CALL IS A CALL THAT INTERCONNECTS THREE OR MORE TELE-
phones, and permits all parties to converse at random. This definition can give
you the feeling that the convenience of connecting three or more telephones
to one line is reserved for companies like IBM and AT&T. A statement like
this cannot be further from the truth.

Figure 13-1 is a schematic of a completely self-contained conference caller
that permits any number of incoming telephone lines to be connected to one
central telephone. The parties connected can enjoy a three-way (or more)
conversation.

The only limitation to this project is the number of incoming telephone lines
provided to your place of residence by the telephone company. Most have only
one. But for the number of hobbyists that exceed the national average, this project can be constructed in one evening. With a little more time in laying out an
attractive housing, you can be rewarded with another useful telephone
enhancement project.

## HOW IT WORKS

By referring to Fig. 13-1, three outside telephone lines designations Line
#1, Line #2, and Line #3 are connected by the user in any sequence to allow
multiple conversations. As an added feature, I included a series combination

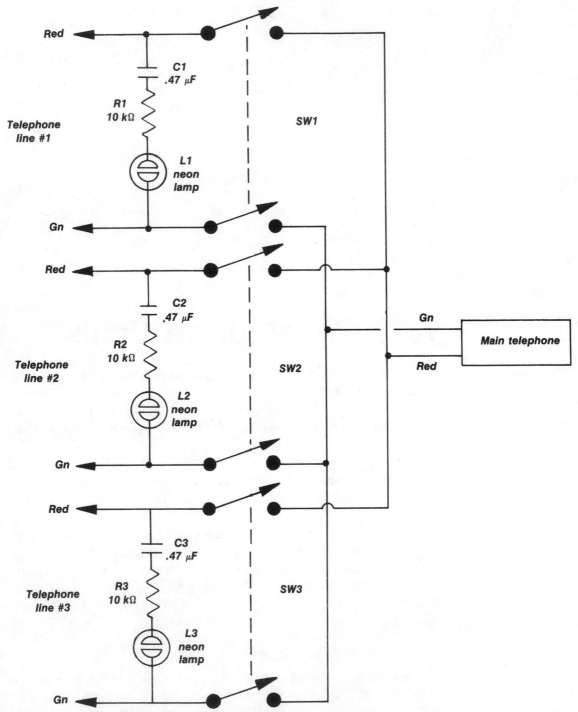

Fig. 13-1. Schematic of the Conference Caller.

**Parts List**

| R1  R2  R3 | 10K Ohm Resistor |
|---|---|
| C1  C2  C3 | .47μf 250V Capacitor |
| L1  L2  L3 | Neon Lamp |
| SW1  SW2  SW3 | DPST (Double Pole Single Throw) Toggle or Slide Switch |
| Misc.  Parts | |
| 1 | Main Telephone |
| 2 or more | Incoming Telephone Lines |
| 2 or more | Telephone Line Cords |
| 1 | Housing |

Fig. 13-2. Parts list for the Conference Caller project.

of a capacitor, resistor, and a neon lamp connected across each incoming telephone line. This combination serves as a visual indicator to provide a means of showing the user which of the three lines is ringing at any one time.

Figure 13-1 shows three incoming telephone lines, but you're not prevented from modifying this circuit to suit the number of lines available at your home or office. If more then three lines are to be interconnected, just repeat the wire configuration for each addition.

Three DPST (Double Pole Single Throw) SW1, SW2, and SW3 toggle or slide switches connect the tip and ring of the desired telephone line to one common connection, where a standard telephone is wired.

With a ringing signal at line one, the neon lamp associated with this line begins to flash. To answer the call, flip SW1 down. This connects the tip and ring of line #1 to the main telephone answered in the usual manner. If, while talking with a party on line #1, the neon lamp of line #2 flashes, just flip SW2 down. This connects line #2 to the previously connected line #1. At this time, a three-way conversation can now take place. After completing the call, just flip SW1, and SW2 back into their normally open (or up) position (this position is shown in Fig. 13-1). This will disconnect the main telephone from incoming lines #1, and #2.

## ASSEMBLY

The conference caller is not a complicated circuit, but care must be taken while wiring the tip and ring (red and green) wires for each telephone line. Maintain proper line-polarity at all times. Don't connect a red wire to a green. This will provide a polarity reversal on the line, and might prevent an older type tone-dial telephone from operating.

If a polarity reversal is made on a standard rotary telephone, no dialing problems will be encountered. Rotary dial telephones like the ones pictured in Chapter 1 do not require a specific polarity.

## INSTALLATION

When the conference caller is complete, check for any unwanted line reversals as mentioned. When everything checks out, begin connecting each incoming telephone line to its associated toggle switch. Again, observe polarity of the line.

To ensure correct installation, wire *one* line at a time and test each line as it is connected, by flipping the toggle switch down, then lifting the main telephone handset. If all checks out, dial-tone should be heard. If a tone-dial telephone is used (2500 or 3554, 2554 type phone) test for correct line polarity by pressing one of the numbered dialing buttons. The normal tone-dial frequencies should be heard. If not, recheck the wiring. When all lines are connected to the conference caller, and all checked for polarity reversal, you can now make conference telephone calls, just like an executive officer at AT&T.

# 14    Telephone Lock

THE HIGH COST OF TELEPHONE BILLS, GENERATED BY THE UNAUTHORIZED USE of business or home telephones can be staggering. Your only defense against such abuse is to install a Telephone Lock. Not one of those cheap series circuit locks using three rotary switches but an elaborate, yet simple, combination system.

The telephone lock presented in this chapter is a two integrated circuit assembly that allows the user to program any four number code using a 4 or 12 button calculator-type keyboard as its input. It is recommended that 12 switches be used as the input because increasing the number of buttons also increases the number of possible combinations you must use in order to find the one code that will allow access to an outside line.

## HOW IT WORKS

The heart of the telephone lock is the 7474 integrated circuit seen in Fig. 14-1. The 7474 is a D type flip-flop circuit wired as a sequential pass-on. The term *sequential pass-on* is used to indicate a state where, in order to get an output a series of events must occur in one specific order. Note SW2 to SW5 in Fig. 14-1. For the X1 relay to pull in, switch A must be pressed *first*, followed by switch B, then switch C, and finally, switch D. Any other sequence will prevent the relay from energizing, preventing access to the telephone dial tone.

104

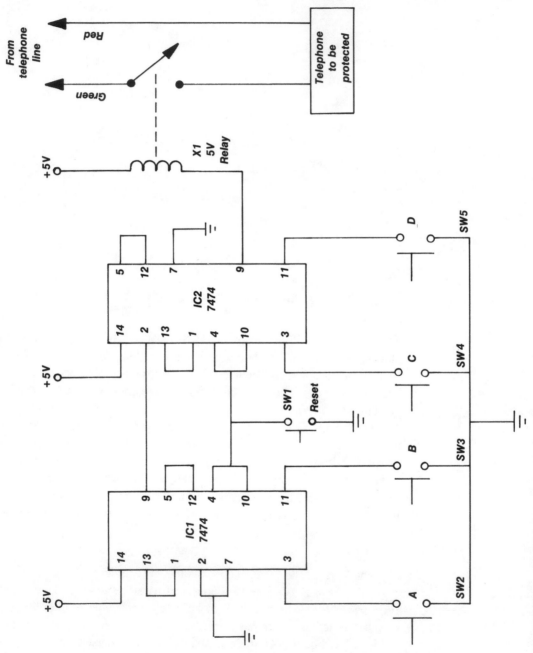

Fig. 14-1. Schematic of the Telephone Lock.

**Parts List**

| | |
|---|---|
| SW1 to SW5 | Spring Return Push Button Switches |
| IC1 IC2 | 7474 D Type Flip-Flop |
| X1 | 5 volt Relay |
| Misc. Parts | |
| 1 | Telephone Line Cord |
| 1 | Rotary or Tone Dial Telephone to be Protected |

SW2 to SW5 can also be a Calculator type keyboard so to increase the possible number combinations.

Fig. 14-2. Parts list for the Telephone Lock.

Only with the proper sequence of switch closings, will relay X1 pull in, connecting the red or green line-cord wire to the telephone being secured. In reality, the contacts of the X1 relay (a small inexpensive 5-volt reed relay can be used as X1) acts as an on/off switch under the control of the keyboard code input.

When the telephone conversation is complete, press SW1 (reset) to allow the relay to drop out again, disconnecting the telephone from the line.

## HOW THE ICS WORK

With the reset button (SW1) pressed, pin 5 of IC1, and pin 9 of IC2, go high (a high indicates a voltage of approximately 3 to 4 volts). When SW2 is pressed, a ground is placed on pin 3 of IC1. This ground changes the state of pin 5 (IC1) from high to ground (0 volts). This 0 volts is passed on to the second D flip-flop also contained in the same IC1 package.

By pressing SW3, a second ground is placed now on pin 11 (IC1). The ground (or 0 volts) on pin 11 changes the state of pin 9 from high to ground. This pin 9 ground is then brought out to pin 2 on IC2, where the same sequence of events takes place now in IC2 by pressing switches SW4 and SW5.

Each flip-flip will change state (from high to low or ground) only when the output of the preceding flip-flip stage is low or ground. If it is not, the flip-flop will not change state preventing the next stage from operating.

## ASSEMBLY

The building of this project is a little more complicated compared to the previous telephone projects, but don't let its complexity prevent you from its assembly. Just remember that pin 1 on all ICs is determined by a small dot on the surface of the plastic package. From this pin, count *counter-clockwise* (see Fig. 14-3).

When soldering ICs, remember that excess heat can destroy the sensitive micro-circuitry, so *do not* overheat each leg when soldering. Consider reviewing

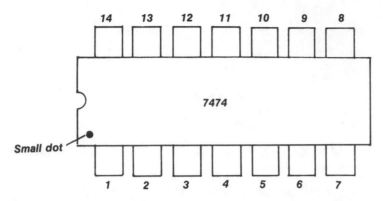

Fig. 14-3. How to count the pins on an IC.

Chapter 5 before attempting the assembly of the telephone lock. This chapter will refresh your memory in the proper soldering techniques and precautions needed when soldering heat-sensitive components.

## KEYBOARD WIRING TECHNIQUES

Figure 14-4 is the recommended keyboard that can be used to input the proper security code to the Telephone Lock. The twelve normally-open pushbutton switches can be mounted on a plastic box in a configuration of a telephone tone-dial.

Note in Fig. 14-4 that one side of all 12 switches is shorted together with the three vertical buss wires as shown. Solder an additional horizontal bus to interconnect the vertical, then solder a short wire to the common ground (pin 7) of the 7474 ICs. Now is the time to determine the security code to unlock your telephone. From Fig. 14-4, the four outside switches will be the code for this example.

Solder a wire to the remaining switch terminal on button #1. Then connect the other end to pin 3 of IC1. This switch will replace SW2 as shown in Fig. 14-1. Do the same for button #3, the * button and the # button, then connect these switches to their corresponding IC pins as indicated in Fig. 14-4.

With this example, to unlock a secured telephone, the code **1,*,#,3** must be entered by the user in the sequence presented, otherwise dial-tone will not be heard on the telephone. This is by no means the only code that can be used. You can make up any combination desired. The only thing to keep in mind is that a number can be used only once in the security code. An example of a bad code is 1,4,2,1 and 5,5,9,0. Note that the number 1 is used two times in the first example while the number 5 is used twice in the second.

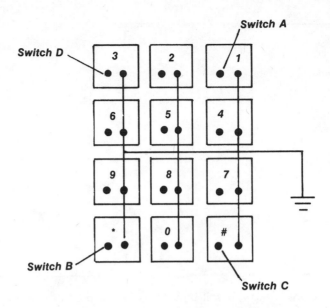

Fig. 14-4. Recommended keyboard wiring for the Telephone Lock.

## POWER REQUIREMENTS

A source of power is required by the lock, this can be in the form of a 9 volt battery. But the ICs require 5 volts for their operation. To drop the 9 volts from the battery to the needed 5 volts, a Zener diode will be used in a circuit called a voltage divider. Figure 14-6 is a voltage divider that can be used with the Telephone Lock.

D1 is a special diode called a zener diode. Its internal characteristics allow a pre-determined voltage to be developed even if the input voltage is larger than the output. For the proper operation of the Telephone Lock, a zener with a rating of 5.1 volts must be purchased. As with an LED, a series dropping resistor must be placed in the circuit. The value of this resistor (R1) depends on the amount of current needed by the two 7474 integrated circuits.

If a databook indicating Integrated Circuit characteristics is available, look up the 7474 and note its current requirements. My book indicates the required current is 17 milliamperes (or 17 mA) per package. Because two ICs are used, the total current required by the project is 34mA.

To allow 5.1 volts at a current rating of 34mA, the value of R1 must be 100 ohms at ½ watt. In Fig. 14-5, this 5.1 volts is taken directly across the Zener diode (D1). Connect a red wire to the + voltage output, and connect a black wire to the − . These two wires are then connected to the Telephone Lock's ICs, as indicated on the schematic. Observe proper polarity when wiring up the power leads (red and black).

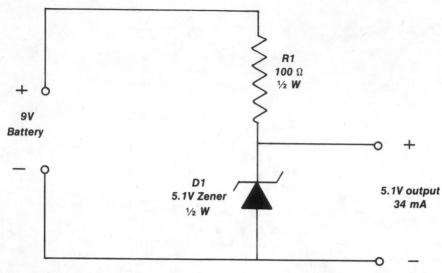

Fig. 14-5. Converting a 9-volt battery to deliver 5.1 volts at 34 mA using a zener diode.

*Parts List*

| | |
|---|---|
| R1 | 100 Ohm ½ Watt Resistor |
| D1 | 5.1 volt ½ Watt Zener Diode |
| Misc. Parts | |
| 1 | 9 volt Battery |

Fig. 14-6. Parts list for the Telephone Lock power supply.

## INSTALLATION

When the lock assembly is complete, determine which telephone is to be secured. The only wiring to be made is either the red or green wire (tip or ring) of the telephone line-cord must have the normally open relay contacts placed in series with it, to act as the required on/off switch. This can be done at the junction box, or if you wish, inside the telephone.

## JUNCTION BOX WIRING

Remove the cover of the junction box with a screwdriver. Locate either the red or green telephone line-cord wire, and remove it from its terminal screw. Connect this wire to one of the X1 relay contacts. Connect the second X1 contact to the terminal screw where the line-cord lead was removed. Installation is now complete. Replace junction box cover.

## INTERNAL TELEPHONE WIRING

Remove the telephone housing and locate the red or green dial-tone wires. These wires are connected to the telephone network at terminal L1 and L2. Unscrew either the red or green wire and connect this lead to one of the X1 relay contacts. Now connect the second X1 contact to the terminal the lead was just removed from. Installation is now complete. Replace the telephone housing.

## HOW TO USE YOUR NEW TELEPHONE LOCK

If a call is to be made from a secured telephone, first lift the handset. At this time, there should be no dial-tone. Using the keyboard, enter the security code the lock was wired for (in Fig. 14-4, this code will be **1,*,#,3**). Upon the entry of the last code number, dial-tone should be heard in the handset. When the call is complete, hang up the handset, then press the reset button. This action will force the X1 relay to drop out, disconnecting the secured telephone from the line. No other calls can be made until the security code is re-entered.

# 15    Telephone Intercom

INTERCOM SYSTEMS HAVE BEEN PART OF INTERNATIONAL BUSINESS ORGANIZA-
tions for many years. This communication technology has not been widely used
in the home because an average home is considered relatively small compared
to multi-million dollar corporate headquarters. The project presented in this
chapter, I hope will change this outlook.

Designed strictly for the home, this ten station, tone-dialing, intercom
system, can provide room-to-room communications at a fraction of the cost of
a commercially assembled unit. The heart of this project is the tone-dial
telephones that must be purchased. An affordable way to purchase a number
of 2500, 3554, or even the 2554 type is to visit one or more flea-markets in
the area. In New York, you can find any type of telephone desired just by walking
down the aisle. Most of all, the price of a 2500 telephone can be only $5.00,
compared to a $45.00 price tag for the same phone if it were new.

A flea market can be your supplier of other electronic goodies at a fraction
of the original price. The schematics illustrate an intercom system with a
maximum of 10 tone-telephones. If this number of phones is too high, by all
means make the necessary parts list corrections so that you do not purchase
additional material that will not be used.

## HOW IT WORKS

The Telephone Intercom will make use of terminology that might be unfamiliar to you. If this is the case, I recommend that you review Chapter 1 (Telephone Basics). With the concept of telephone basics fully understood, examine how the intercom operates.

All the tone-telephones purchased have been wired in parallel (as seen in Fig. 15-1). Resistors R1 and R2 supply a voltage of 24 volts to all telephones. This voltage will allow any telephone to supply a side tone. This electrical signal can be heard as sound, when any other telephone in the systems handset is lifted.

With this type of wiring, there is no privacy. Any third-party can lift a handset and enter into the conversation. But for a home telephone system, this parallel wiring is more adequate. To call into extension telephone #5, simply pick up any telephone handset and press the number 5 button. When pressed, the special dialing tones are generated, and can be heard in the receiver.

These tones can be found at the audio transformer T1. The capacitor C1 blocks the talk battery (24 volt power-supply) voltage from coming into the transformer, but allows the audio tones generated by the telephone dial to pass. Through magnetic coupling, the tone-dial output is transferred to the secondary windings of T1, where the input of IC1 is connected. IC1 is a special Tone-Dial Decoder that can be purchased from Del.Phone Industries. (Their address can be found in the parts list). This M-957-01 IC transforms the Tone-Dial's audio output back into the corresponding number that has been dialed. This output is applied to pins 1,20,21, and 22 as a binary code output (binary code is a computerized format that allows you to represent numbers or letters as a series of 1's and 0's).

To further decode the output of IC1, the binary code is applied to the input of IC3, (see Fig. 15-2) a 1 of 16 data distributor. This IC is used to provide a 1-low-out-of-16 output when the corresponding binary input is present. When there is no tone-dial frequency present, all the output pins of IC3 (pins 1 to 10) have a high (or positive voltage) on them. When a telephone handset is lifted and the #5 button is pressed on the tone-dial, the corresponding binary code is inputted to IC3 by the tone decoder (IC1). At this time, IC3 will convert this binary code back into the number 5, by allowing pin 5 to go low (or 0 voltage) while all other output pins will remain high.

IC3 is a 1-of-16 distributor, however, the telephone intercom makes use of ten lines, the remaining six IC3 outputs are not used, and will remain unconnected. Now that IC3 has converted the tone-dial frequencies back into the number 5, a means of "ringing" extension #5 has to be incorporated at this time. Relays X1 to X10 will provide this function.

With no binary code present at the input of IC3, all output pins will be high, preventing any of the 10 relays from pulling in, but with the binary code for the number 5 decoded by IC3, pin 5 will go low (or ground) allowing the relay

113

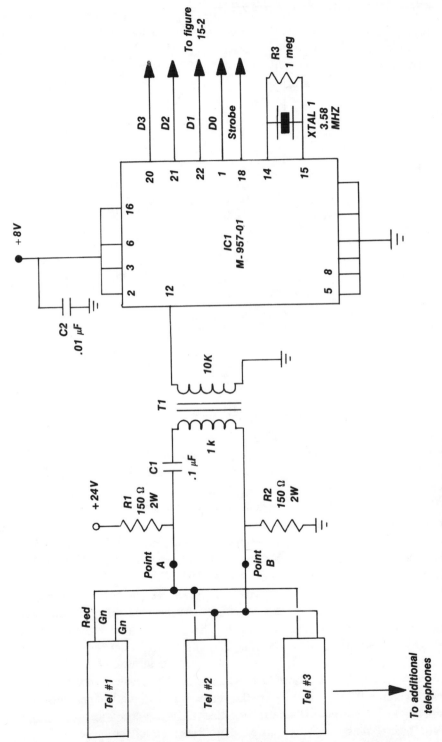

Fig. 15-1. Schematic of the Telephone Intercom.

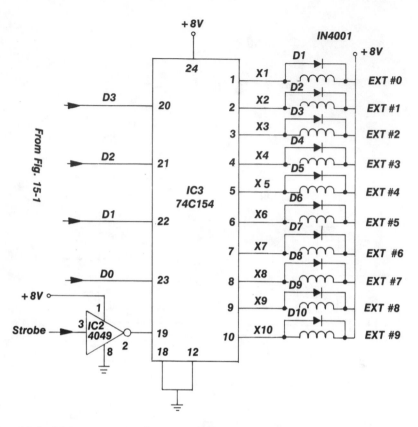

Fig. 15-2. More of the Telephone Intercom.

X5 to pull in. This relay will remain energized for as long as the number 5 button is pressed at the telephone end. The same holds true for any other dialed extension number.

At this time, you must decide how to signal the called telephone. This circuit allows two options. One being the builder purchase 12-volt ac buzzers. These buzzers can be purchased at any electronic mail-order house for about $1.00. The second option being, the hobbyist assemble an audio oscillator. This oscillator will produce a tone whenever voltage is applied.

At this point in our example, if you wish to use a 12-volt buzzer as a ringing device, Fig. 15-3 shows the interconnecting wiring for the 10 relay contacts. One side of all relay contacts is common. This is the point where the ringing voltage is to be applied. Whether it be 12 volts ac or 8 volts dc (but can not be both for obvious reasons). Let's go back to the point when the X5 relay is energized, from Fig. 15-3, note that the contacts of X5 close. This closing will apply a voltage (12 Vac or 8 Vdc) to the yellow wire in the telephone line-cord (note point C). Now, inside the telephone, this yellow line-cord wire is connected to the buzzer or the audio oscillator. The black wire of the line-cord is the

115

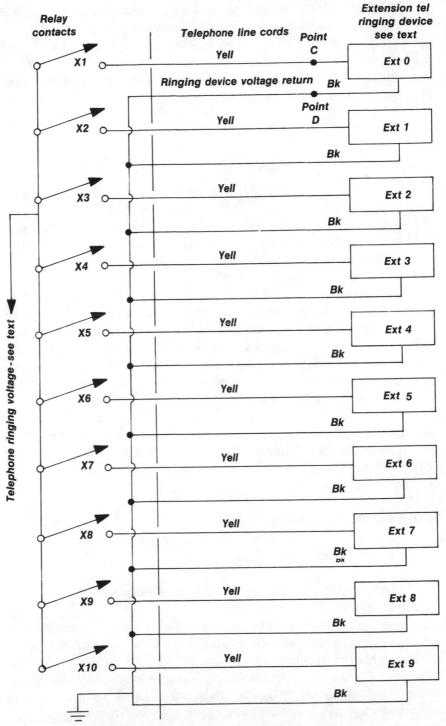

Fig. 15-3. How to wire the ring relays for the intercom.

common return (note point D). This black wire is brought back to the main intercom electronics where it is connected to ground.

With the electrical loop now completed, the ringing device will be turned on every time the contacts of X5 close. Obviously, this explanation will hold true with any relay being energized. With extension #5 now ringing, the called party just lifts the telephone handset allowing normal conversations to take place. When complete, just return the handset to its cradle.

## POWER SUPPLY

The power supply for the Telephone Intercom looks rather complicated, but really is not. An ac voltage of 117 volts is applied to the primary windings of both the T2 and T3 transformers. The output of T2 is 24 Vac and the output of T3 is 12 Vac. Both ac voltages are rectified into pulsating dc by the action of the bridge rectifiers (D11 to D14 at the secondary of transformer T2, and D15 to D18 at the secondary of transformer T3).

The pulsating dc is then filtered by capacitors C3 and C4. This filtering transforms the pulsing dc into a relatively flat dc signal. To filter this dc even further, voltage regulators are used, REG1 is a 7824T regulator that provides a pure 24 volt dc signal output used to supply the talk battery for the telephones (this voltage is applied through resistor R1), REG2 (7808) is also a voltage regulator, but its output is 8 volts compared to the 24 volt output for REG1.

This 8 volts dc is applied to pins 2,3,6, and 16 of IC1, pin 1 of IC2, pin 24 of IC3, and as a common applied-voltage to relays X1 to X10. This output can also be used to supply the voltage needed to operate the audio oscillator (ringing device) in each telephone. The ac voltage used if a buzzer-type ringer device is decided upon, can be obtained at the secondary of transformer T3 (see Fig. 15-4).

CAUTION: This power supply makes use of the 117 Vac. If this wiring is unfamiliar to you, do not build this power supply. Use batteries to supply the required dc voltages.

To further explain the wiring required by each telephone, please refer to Fig. 15-6.

The normally used red and green wires of the line-cord still supply the talk battery to the telephone network so that a side-tone can be generated as if it were connected to an outside telephone line. The additional two wires in the line-cord (yellow and black) provide a ringing voltage (yellow provides the ringing voltage) and a ground return (ground return is provided by the black wire). In a standard telephone, the yellow and black wires are usually not used, so they are taped and stored.

117

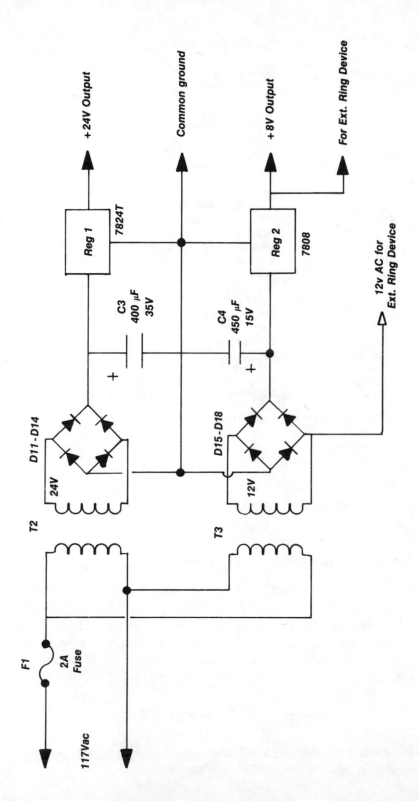

Fig. 15-4. Suggested power supply for the intercom.

*Parts List*

| | |
|---|---|
| R1 R2 | 150 Ohm 2 Watt Resistor |
| R3 | 1 Megohm Resistor |
| C1 | .1$\mu$f Capacitor |
| C2 | .01$\mu$f Capacitor |
| C3 | 4000$\mu$f 35 volt Capacitor |
| C4 | 450$\mu$f 15 volt Capacitor |
| T1 | Audio Transformer   1K Ohm to 10K Ohm |
| T2 | Power Transformer 117Vac Primary 24Vac Secondary |
| T3 | Power Transformer 117Vac Primary 12Vac Secondary |
| XTAL1 | 3.58 MHz. Crystal |
| Reg1 | 7824T Voltage Regulator (24V) |
| Reg2 | 7808 Voltage Regulator (8V) |
| IC1 | M-957-01 Tone Dial Decoder |
| IC2 | 4049 IC |
| IC3 | 74C154 IC |
| D1 to D18 | 1N4001 Diode |
| X1 to X10 | SPST (Single Pole Single Throw) 6 volt dc Relay |
| F1 | 2 Amp Fuse |
| Misc. Parts | |
| 10 or less | Tone Dial Telephones |
| 10 or less | Telephone ringing device for mounting inside tone phone |

IC1 can be purchased from:

Del. Phone Industries
4487 Plumosa St.
Spring Hill, FL 34606 (after Feb. 1, 1989)

Fig. 15-5. Parts list for the Telephone Intercom.

## TELEPHONE RINGING DEVICE (AUDIO OSCILLATOR)

Figure 15-7 is a typical audio oscillator that can be made using a readily available 4049-type integrated circuit. The audio output is applied to a standard telephone receiver element. This output should provide adequate volume. Note where the yellow and black telephone line-cord wires are connected to the circuit. The frequency of the oscillator can be changed by varying either the R1 resistor or the capacitor C1. The frequency can be calculated by the formula:

$$\text{Frequency} = 1/(2.2 \times \text{R1} \times \text{C1})$$

Before installation inside the telephones, test each oscillator for proper operation. This will prevent additional troubleshooting later on.

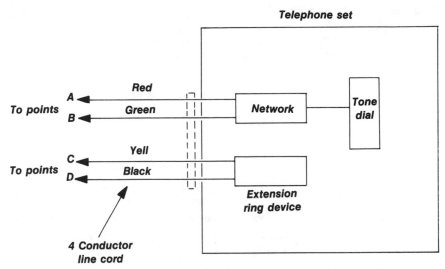

Fig. 15-6. How to wire a telephone for side tone and for ringing.

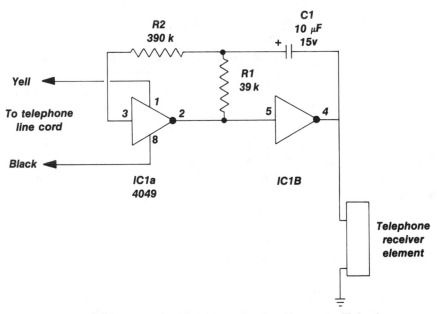

Fig. 15-7. A suggested tone generator for the Intercom Telephones.

## INSTALLATION

Before installation, test the operation of the intercom by temporarily connecting a telephone to each relay and talk battery, checking for proper decoding of each tone-dial button, and for proper ringing. When all seems to

**120**

*Parts List*

| | |
|---|---|
| R1 | 39K Ohms Resistor |
| R2 | 390 K Ohms Resistor |
| C1 | 10µf 15 volt Capacitor |
| IC1 | 4049 IC |
| Misc. Parts | |
| 1 | Telephone receiver element |

Fig. 15-8. Parts list for the Tone Generator circuit.

function, determine where the telephones are to be located. Using four-conductor cable, run the cable from each telephone to one central point where the intercom electronics will be located. Remember to place a piece of tape (flagging) on each cable so the location of each telephone can be determined. Connect all telephone red wires together and connect all green wires together. Take the common red wire and fasten it to point A of the intercom, then fasten the common green wire to point B (see Fig. 15-1 for the location of these two points).

When complete, lift the handset of each telephone and check for side-tone. Also check the tone-dial for an output to make sure no polarity reversals have taken place. Connect all telephone line-cord black wires together and fasten to the common ground of the intercom.

It's now time to wire up the telephone ringer. But first, determine the telephone extension number for each location and note it somewhere. If the telephone in the bedroom is to be extension #1, connect its yellow wire to the X2 contact (see the importance of labeling the cables when installing the phone?). If a phone in the basement is to be extension #0 (or Operator), connect its yellow wire to the X1 relay contact. Do the same for the additional phones in the system.

With the Telephone Intercom plugged-in (or operating from batteries) lift a telephone handset and check for side-tone. Dial the 10 extension numbers to check for proper decoding and ringing. When all is working, it's time to enjoy the convenience of your very own telephone intercom system.

# 16 Telephone Line Tester

WITH THE AMOUNT OF INFORMATION SUPPLIED IN CHAPTER 1 (TELEPHONE BASICS) anyone who is handy with a few common tools will be able to install their own extension telephones, telephone junction boxes and enhancement equipment. With all this time invested, you will want to quickly and accurately test each newly installed modular junction box as well as the cable run and line polarity. With the circuit presented in Fig. 16-1 you can do just that.

The Telephone Line Tester is not only a fast way to check if dial-tone is present, it will also check for telephone line-polarity reversals. It is these reversals that prevent the older type of tone-dial telephones from operating correctly. And if a call is being received, just press SW1 for a visual indication of the ringing signal. The voltage of this ac can also be determined by the brightness of the light. All this can be easily built for under $3.00 (excluding housing).

## HOW IT WORKS

The dc voltage on a telephone line is about 50 volts. A dc voltage unlike the ring signal that has alternating positive and negative cycles every $1/30$ of a second, has a positive voltage on the green wire while a negative voltage is on the red. The presence of this dc voltage is quite easy to detect. By wiring

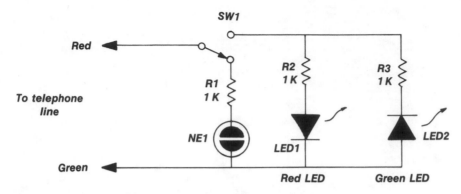

Fig. 16-1. Schematic of the Telephone Line Tester.

two LED's across the telephone line with a dropping resistor in series with each, you can have a visual indication of the voltage and the polarity. If the red LED lights the polarity of the line is reversed and should be corrected. If the green LED lights this indicates two important factors, the first being that the polarity of the line is correct, and if a short one-time flash is noticed on the green LED, this indicates that dial-tone is present.

For two LED's to detect polarity changes on a dc line, one LED must be installed in reverse to the other. Note this reversal in Fig. 16-1. If neither LED lights when connected to the telephone line, this indicates an open wire or bad connection somewhere in the cable itself, or at the place of splicing. For the LED's to be placed across the line, (to test for voltage and polarity), press SW1.

To allow for the testing of the 120 Vac 30 Hz ringing signal, a neon lamp with a dropping resistor (R1) will be used. As discussed with the telephone call indicator project (Chapter 11) a neon lamp can make a great ring detector because the bulb requires at least 90 volts to produce light. The amount of light can indicate to the user if the ringing voltage is high enough to activate the clapper in the bell so that an audible sound is produced.

**Parts List**

| | |
|---|---|
| R1  R2  R3 | 1000 Ohm Resistors |
| NE1 | Neon Lamp |
| LED1 | Red Light Emitting Diode |
| LED2 | Green Light Emitting Diode |
| SW1 | SPDT Slide or Toggle Switch (Single Pole Double Throw) |
| Misc. Parts | |
| 1 | Telephone Line Cord |

Fig. 16-2. Parts list for the Line Tester.

If a ringing voltage is known to be present, do not press SW1. Resistor R1 and NE1 are already connected to the line through the action of SW1. If the neon lamp (NE1) lights to a normal (what is normal will require experience on the part of the user) brilliance, the voltage on the line will be high enough to ring the internal telephone bell.

If no light is generated by NE1, this does not mean that there is no voltage, but it does indicate that the voltage on the line is below 90 volts. Any ringing signal below 90 volts will not ring the telephone bell, and if it does, the bell will not sound right. If this is the case, there might be a drain on the line. Another telephone bell on the line might have a shorted ringer coil. This will draw an abnormally high amount of current preventing the other extensions from ringing. To find the problem, disconnect each extension telephone, one at a time, until there is a normal flashing on NE1 of the tester.

The component SW1 can be a toggle, slide, or push-button switch (I recommend a push-button type switch), depending on the requirements of the user. SW1 is called a SPDT (single pole double throw) spring return switch or push button. This switch can connect two different circuits to one common point. (In the case of the telephone line tester, the common point is the red telephone line-cord.) Without pressing the button, the internal contacts are positioned as shown in Fig. 16-1. This allows the neon lamp and resistor R1 to be connected across the telephone line. This position tests for the incoming ringing signal. At this time, the two LED's are disconnected from the line, protecting them from the high ringing voltage. If SW1 is pressed, the red line-cord wire will now be connected to the two LED's. In this position, you can test for line voltage and polarity reversals.

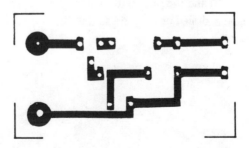

*Telephone tester*

Fig. 16-3. PC board artwork for the Line Tester.

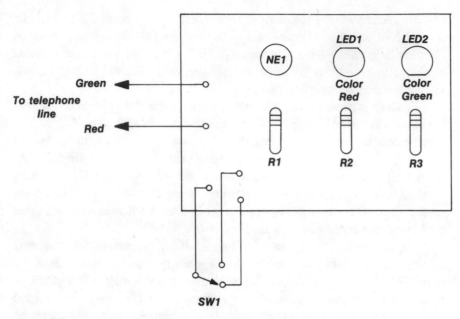

Fig. 16-4. Component layout for the Line Tester's PC board.

## ASSEMBLY

From Fig. 16-1, the Telephone Line-Tester is not complicated to construct. To help in the assembly, printed circuit board artwork for the tester is presented in Fig. 16-3. While the component layout is in Fig. 16-4. Making use of the PC board artwork will guarantee a working project. If the etching of a PC board is too complicated, the famous perforated board can also be used.

The tester should be installed in a small plastic box with the three lamps (NE1, LED1, and LED2) and the push button switch (SW1) protruding from the top. While a small hole is drilled on the side so that the test leads (red and green) can be poked through.

# 17 Telephone Amplifier

EVER GET INTO A SITUATION WHERE A LARGE GROUP OF PEOPLE WANT TO LISTEN into a telephone conversation? How about the time when the kids want to say hello to grandma who lives in Florida? The Telephone Amplifier, in Fig. 17-1 will surely rectify this situation without a large dollar investment on the part of the builder.

Unlike the other enhancement projects listed in this book, the telephone amplifier is not connected directly to the telephone line but rather is wired to the telephone network (a small printed circuit board or metal box which contains the telephone's electronics), and with a small power amplifier (LM380) this project provides room filling sound.

## HOW IT WORKS

By referring back to the schematic of a standard 500 (rotary desk) telephone (Fig. 1-24), notice that the receiver element of the handset is connected to the telephones electronics (network) using two white wires from the coil cord. One being connected to terminal "GN" and the second to terminal "R". At these two points an audio signal of the distant party is produced along with a side-tone.

The *impedance*. (The total opposition of a circuit to the flow of an audio or ac signal) between the "GN" and "R" terminals is quite high. This impedance is in the order of 2000 ohms. This is the reason that the telephone receiver

126

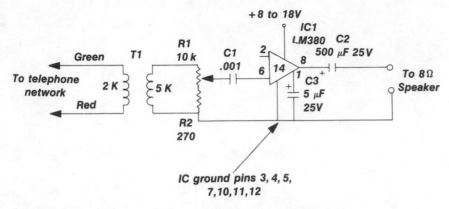

Fig. 17-1. Schematic of the Telephone Amplifier circuit.

**Parts List**

| | |
|---|---|
| R1 | 10K Ohm Potentiometer |
| R2 | 270 Ohm Resistor |
| C1 | .001μf Capacitor |
| C2 | 500μf 25V Capacitor |
| C3 | 5μf 25V Capacitor |
| T1 | Audio Transformer 2K Ohm to 5K Ohm |
| IC1 | LM380 Power Audio Amplifier |
| Misc. Parts | |
| 1 | 8 Ohm Speaker |
| 1 | 8 to 18 volt battery or power supply |

Fig. 17-2. Parts list for the Telephone Amplifier.

is of a high-impedance crystal-type. To increase the audio sound from this point, an amplifier can be connected in parallel with the receiver element. The impedance of an amplifier IC is even greater than the crystal receiver itself. To prevent losing any volume due to a mis-match in impedance between terminals "GN" and "R", and the audio amplifier (network terminals impedance equals 2000 while the impedance of an amplifier input is equal to 5000 ohms. This is a mis-match of 3000 ohms), an audio transformer is connected between these two points. Transformer T1 will match the unbalanced impedances so that the maximum audio signal can be applied to the amplifiers input.

Resistor R1 (10 k Ω potentiometer) is used as the main volume control. This control should be mounted on the front panel of the intended housing. The resistor R2 (270 ohms) is installed in the circuit to prevent a setting on R1 that will eliminate the output from the amplifier altogether. Even at its lowest setting a very low audio signal is still present on the speaker because R2 prevents the signal from being completely shunted (shorted) to ground by introducing a 270 ohm resistance.

Once the desired volume is set by R1, the audio proceeds to the input (pin 6 of IC1) of the power amplifier. This amplifier is capable of producing a two watt audio signal when an 18 Vdc voltage is applied. This two watts is more than enough to produce a room filled with sound. Unlike older tube type amplifiers that require a second audio transformer (audio output transformer) for impedance matching, the LM380 provides an output (pin 8) that is compatible with a standard 8-ohm speaker. The only requirement is that a 500 $\mu$F 25 V capacitor be installed between the two.

The Telephone Amplifier requires an outboard power supply. This voltage can range from as little as 8 volts to a maximum of 18 volts. This power should be free from 60 Hz ''Buzz'' (this affect can also be called ripple). This means that the power supply should contain a voltage regulator. In Chapter 18, I included the schematic (Fig. 18-3) of a fully regulated 12-volt power supply. This circuit can be built using the PC board artwork (see Fig. 18-5) and the component layout (Fig. 18-6).

For the required ac voltage input, I strongly recommend the use of a UL approved wall transformer. This type of power transformer has no high voltage points that can be accidentally touched. All coil windings are encased in a plastic housing that can be plugged directly into a wall socket. Unlike using a chassis-mounted transformer which can produce a nasty shock if not carefully isolated and wired. For an added convenience, you might consider purchasing the 10 k $\Omega$ potentiometer (R1) with an on/off switch, so that the power to the IC can be cut off when not in use.

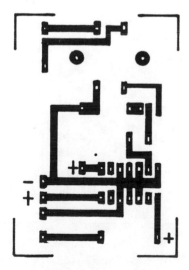

*Telephone amplifier*

Fig. 17-3. PC board artwork for the Telephone Amplifier.

## ASSEMBLY

The assembly of the Telephone Amplifier is straightforward and will not cause any problems. Just remember to keep the wired input of IC1 very short. This will prevent any pick-up of unwanted hum. Also remember to use a well regulated power supply. If not, the 60 Hz. buzz or ripple will be heard in the speaker, this can be very annoying.

To prevent the introduction of any ac hum into the circuit, Fig. 17-3 is a complete layout of the circuit that takes into consideration the length of the input lead, or in the case of a PC board, the length of the copper trace. The component layout of the board is illustrated in Fig. 17-4.

## INSTALLATION

When all is complete and tested, remove the housing from the telephone and locate the screw terminals "GN" and "R". These terminals can be found on the network. If these locations cannot be found, just look for the points where the two white telephone coil-cord wires are connected to.

With a length of shielded cable, connect one wire to terminal "GN" and a second to terminal "R". The other end of this cable should be connected to the 2 k $\Omega$ coil winding of transformer T1. When complete, replace the telephone housing. To use, lift the handset and set R1 to the desired listening level. To prevent feedback (a howling produced by having an amplified output too close

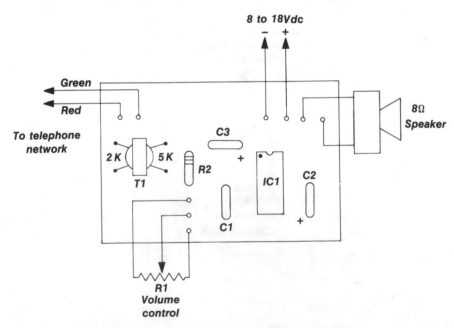

Fig. 17-4. Component layout for the Telephone Amplifier.

to its input source), place the telephone amplifier a few feet away from the main telephone.

This circuit is not an expensive speaker phone unit. This project will not operate if the telephone handset is returned to the cradle. If the telephone is hung-up, the call will be lost. The telephone amplifier can be used only if the handset is removed and kept from its cradle during a conversation. When a call is complete, hang-up the telephone, and if wired into the power supply, turn off the amplifier.

# 18   Speaker Phone

IN CHAPTER 17, I PRESENTED A TELEPHONE AMPLIFIER PROJECT THAT INCREASED the sound output of the receiver to a level that allowed it to drive a speaker. A great project if the builder requires a receive-only amplifier. Chapter 18 takes this idea one step further. By adding a pre-amplifier, one smaller power amplifier, and one more transformer, the Telephone Amplifier can be easily converted to a completely self-contained Speaker Phone.

The sensitivity of the microphone is so high that it will allow the user to be located across a small-size room while conversing, yet still be heard clearly at the other telephone. The speaker phone is an answer only telephone. There is no means of dialing a number, but it can be wired in if so desired. It also lacks a bell or some kind of ringing device. The bell was left out on purpose. With all the enhancement projects in this book, you can easily incorporate one, or even install a device of your own design across the telephone line. The final appearance of the speaker phone depends on the requirements and taste of the user. The designing capabilities of the hobbyist is put to the test. Feel free to experiment and enjoy.

## HOW IT WORKS

Figure 18-1 is the complete speaker phone (except for the required power supply). There is no telephone network incorporated. The two audio trans-

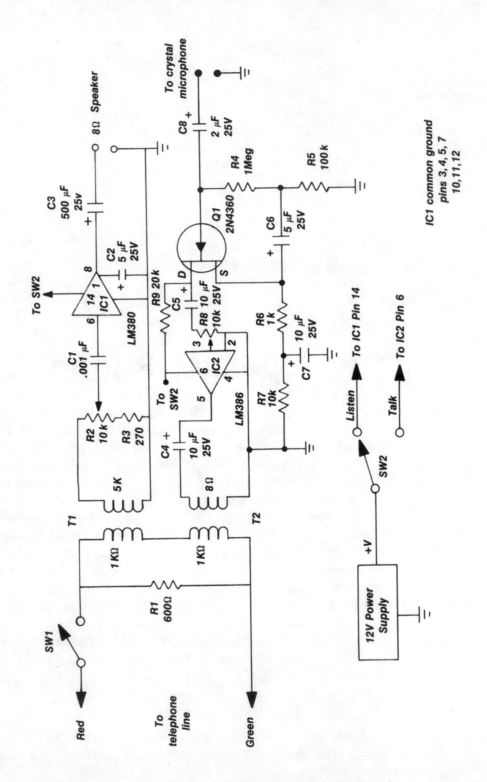

Fig. 18-1. Schematic for the Speaker Phone.

**Parts list**

| | |
|---|---|
| R1 | 600 Ohm Resistor |
| R2 R8 | 10K Ohm Potentiometer |
| R3 | 270 Ohm Resistor |
| R4 | 1 Megohm Resistor |
| R5 | 100K Ohm Resistor |
| R6 | 1K Ohm Resistor |
| R7 | 10K Ohm Resistor |
| R9 | 20k Ohm Resistor |
| C1 | .001$\mu$f Capacitor |
| C2 C6 | 5$\mu$f 25V Capacitor |
| C3 | 500$\mu$f 25V Capacitor |
| C4 C5 C7 | 10$\mu$f 25V Capacitor |
| C8 | 2$\mu$f 25V Capacitor |
| T1 | Audio Transformer 1K to 5K Ohms |
| T2 | Audio Transformer 1K to 8 Ohms |
| IC1 | LM380 IC |
| IC2 | LM386 IC |
| Q1 | 2N4360 |
| SW1 | SPST (Single Pole Single Throw) Toggle or Slide Switch |
| SW2 | SPDT (Single Pole Double Throw) Push Button Switch with spring return |
| Misc. Parts | |
| 1 | Telephone Line Cord |
| 1 | 8 to 18 volt dc Power Supply |

Fig. 18-2. Parts list for the Speaker Phone project.

formers (T1 and T2) are a substitute. Transformer T1 has an impedance of 1000 to 5000 ohms while T2 has an impedance of 1000 to 8 ohms. The primary of both T1 and T2 are wired in series. This wiring will give a total impedance of 2000 ohms. The shunting resistor (R1) is used to provide for the central office, a constant resistance of about 600 ohms. With a parallel combination of 600 ohms and 2000 ohms, the total resistance of the primary telephone line input is about 550 ohms. This resistance will prove to be satisfactory for the design.

The audio signal is applied to the two series-wired coils and transferred to both secondary windings. Transformer T1 will provide an output impedance of 5000 ohms to the audio input of the LM380 IC. The operation of this section is the same as described in Chapter 17.

A crystal microphone is mounted somewhere in the telephone housing, and is used to detect the voice pattern of the user. It is then amplified a small amount (this stage is called a pre-amplifier) by transistor Q1. This is a special transistor called an FET (Field Effect Transistor). The three legs of an FET are named differently than a standard transistor. In an FET device, the legs are called gate, drain, and source. When wiring the circuit, be careful not to confuse the different designations and mount the FET backwards.

When amplified a small amount by the Q1 FET, the audio signal is then applied to a power amplifier (IC2) through a volume control R8 (the setting of R8 is critical, this setting will prevent the audio signal from overloading the telephone line). The output is coupled to the 8-ohm winding of transformer T2, where it is applied to the telephone line.

Both the received, and transmitted audio is applied to both transformers, feedback will result, and an unusable project will result. To compensate for the feedback problem, engineers of ITT and other larger telephone companies have developed expensive and elaborate electronics that permit the receiver and transmitter to be turned on and off only when that particular section is needed.

To eliminate the need for the additional electronics, SW2 was incorporated. This is a SPDT spring-return pushbutton switch that applies power to either pin 4 or IC1 when you wish to listen to the caller, or pin 6 of IC2 when you wish to respond. This switching arrangement is identical to using a walkie-talkie. Remove your finger from the button when you wish to listen. And press the button when you wish to talk.

Figure 18-3 illustrates the recommended power supply for the speaker phone. It delivers a well regulated 12-volt dc output that is then connected to the common arm terminal of SW2, so it can be distributed to either IC1 or IC2.

## ASSEMBLY

Take your time and select components of quality. The speaker phone uses audio frequencies, substitutions in component values can take place, but do not exaggerate by substituting a 5000 ohm resistor for one that is supposed to be 270 ohms. This is a sure-fire way to build a project that refuses to operate.

Another consideration is the same as presented in Chapter 17. Using short interconnecting wires for the input of IC1. Special care must be exercised when wiring the pre-amplifier circuit (Q1 FET). This is a high-gain stage. If not assembled correctly, it will provide a minute amount of hum. This hum will then

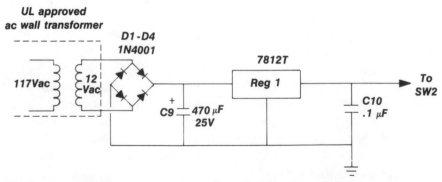

Fig. 18-3. Suggested power supply for the Speaker Phone.

*Parts List*

| D1 to D4 | 1N4001 Diodes |
|---|---|
| C9 | 470$\mu$f 25V Capacitor |
| C10 | .1$\mu$f Capacitor |
| REG1 | 7812T 12 volt Voltage Regulator |
| Misc. Parts | |
| 1 | 12 VAC UL Approved Wall Transformer |

Fig. 18-4. Parts list for the power supply.

be pre-amplified in the power stage (IC2). This hum will be very loud and annoying to the party on the other end.

When complete, connect a well regulated 12-volt power supply to SW2. Connect the speaker phone to the telephone line and flip SW1 down to connect the T1 and T2 transformers to the line. At this time, you will hear a dial-tone coming from the speaker phone. Adjust R2 for the desired listening level. The sound level of a telephone call varies, R2 should be a control mounted on the faceplate of the final telephone housing.

To test the transmitter section of the speaker phone, adjust R8 to its mid-setting. Press SW2, at this time the 12-volt power supply is now connected to IC2 and to R9. While having someone on an extension telephone talk at a normal level and adjust R8 until the voice-level on the extension is adequate. Do not allow too much signal to be applied to the telephone line. This will only develop distortion. Potentiometer R8 is adjusted only once. For this reason R8 should not be mounted on the housing, but rather on the PC or perforated board. This will prevent unauthorized adjustment of this component.

When the call is complete, flip SW1 back into its up position, this will disconnect the speaker phone from the line. You can also consider using a DPDT (double pole double throw) switch for SW1. If used, it can be wired in such a way that when a telephone call is terminated SW1 will also disconnect the 12-volt power from the arm of SW2, turning off the unit.

## ADDITIONAL RECOMMENDATIONS

Figure 18-3 is a schematic of a 12-volt power supply that can be used on the speaker phone, or any other desired project. For the required 12-volts ac, I recommend using a UL approved ac wall transformer. This will prevent any shock hazards from creeping into the circuit, and it will also protect the telephone line from stray ac.

If you wish to etch your own PC board for the power supply, Fig. 18-5 provides the needed artwork. Figure 18-6 shows the component placement. It is recommended that a second power supply be built and enclosed in a plastic

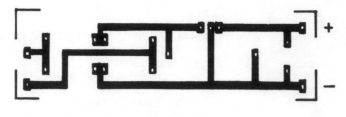

*12 volt power supply*

Fig. 18-5. PC board artwork for the 12-volt power supply used with the Speaker Phone.

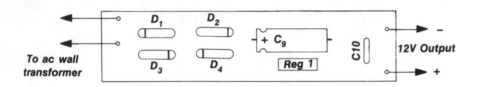

Fig. 18-6. Component layout for the power supply PCB.

box. This second power supply will provide the hobbyist a highly stable supply of 12 volts that can be used on this project or any other project or design that may be of interest.

If you require an easier way to build the speaker phone, Fig. 18-7 will provide the answer. The artwork is the complete layout, including a built-in 12-volt power supply of the project. The artwork is complicated, so take your time when installing the components. Figure 18-8 shows the physical location of each component.

With a PC board of this size, care must be taken when soldering. Do not apply too much solder, if you do, you will run the risk of solder bridges. This is especially true when soldering the two integrated circuits because the spacing of the pins is so close. If care is taken now with the assembly, you will posses a sense of accomplishment and pride every time your telephone enhancement project is used.

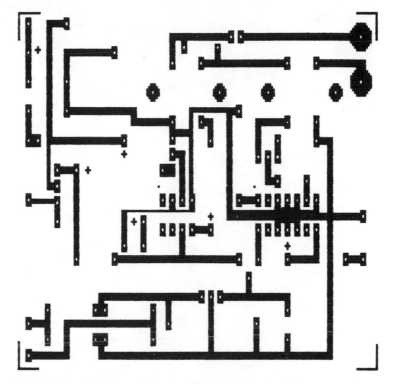

Fig. 18-7. PC board artwork for the Speaker Phone. Note artwork contains a built-in power supply.

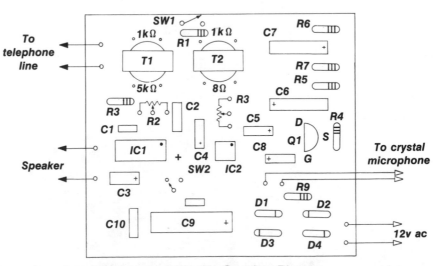

Fig. 18-8. Component layout for the Speaker Phone.

# 19 Appliance Controller

BY MODIFYING THE TELEPHONE INTERCOM OF CHAPTER 15. YOU CAN TURN ON AND off, up to ten appliances using a tone telephone located anywhere in the world. The project in this chapter will not only answer a ringing telephone but allow you to indicate by way of a button on a tone-type telephone, which light in a house to turn on, or you might want to turn on an air conditioner without even being there. This is by far the most useful telephone enhancement device ever published.

## A WORD OF CAUTION

This project makes use of the 117-volt ac house current. If you are not familiar with handling ac—please do not attempt any high-voltage wiring associated with this project. Instead you can use the appliance controller for low voltage equipment.

The Appliance Controller makes use of the M-957-01 and 74C154 Integrated circuits. Instead of having its input (IC1) connected to a telephone intercom, the transformer T1 interfaces pin 12 (IC1) directly to the telephone line, through the normally open contacts of the X1 relay. The X1 relay contacts (designated on the schematic as X1B-X1C) act as a telephone hook-switch under the control of IC2 (4N33 opto-isolator IC).

The circuit containing IC2 is a familiar one. It is taken from the melody ringer project. But instead of controlling a 1.5 Vdc power source for a melody generator, the X1 relay closes, connecting the primary of T1 (1 k Ω) to the telephone line. The 600 ohm resistor (R1) is in parallel with the primary of T1. This is to ensure that the central office switching equipment will always see a constant 500 to 600 ohm load. This resistance is a substitute for a standard telephone.

Using a 120 Vac 30 Hz telephone ringing signal applied to the red and green wires, a small voltage is generated across pins 1 and 2 of IC2. Across these pins is a small internal LED that flashes on a light sensitive Darlington transistor circuit. When no light is shining on the transistors, there is a high resistance between pins 4 and 5, an open circuit. When the internal LED flashes (when a telephone call is being received), the open circuit (between pins 4 and 5 of IC2) is now a low resistance. This low resistance allows ground (negative voltage) to be placed on pin 5. This will allow capacitor C4 (330 µF) to charge up and also turn on transistor Q1.

When Q1 starts to conduct relay X1 closes. As stated before, X1 relay contacts X1B and X1C close, connecting the controller to the telephone line. Notice relay contacts X1A. This set of contacts convert a standard relay into a latching relay. Note from the schematic at Fig. 19-1, when transistor Q1 conducts, the collector leg of Q1 is now at +8 volts. This voltage is applied to one side of relay X1. The other side is connected to ground. With this arrangement, X1 pulls-in, thus closing all four (note only three are being used in this circuit) contacts. To keep the X1 relay closed even when there is no ringing voltage, an outside +8 volts has to be introduced at the collector of transistor Q1 when it is not conducting. This +8 volts is delivered to the collector of Q1 by the contacts of relay X1 (contacts X1A). When contacts X1A close, a +8 volts is applied to relay X1 through the normally-closed contacts of relay X2 (more on relay X2 in a moment). This positive voltage on one side of the relay, and a ground on the other side will keep the X1 relay pulled in until it is time for the controller to be disconnected from the telephone line (this is also under the control of the user).

Now with the controller on line, the user can now push a button on the telephone tone-dial. These special frequencies ride on the telephone line until it reaches pin 12 of the tone decoder chip (IC1) where the tones are converted back into that specific number pushed by the user. This conversion takes place in IC1 and IC4. Remember how tones are converted back into the selected number? If not, refresh your memory by flipping back and re-reading this section of Chapter 15.

For an example, say the user accesses the controller and presses the number five tone-dial button. At the receiver end, these tone-dial frequencies are converted back into the number five. This conversion can be seen at pin 6 of

141

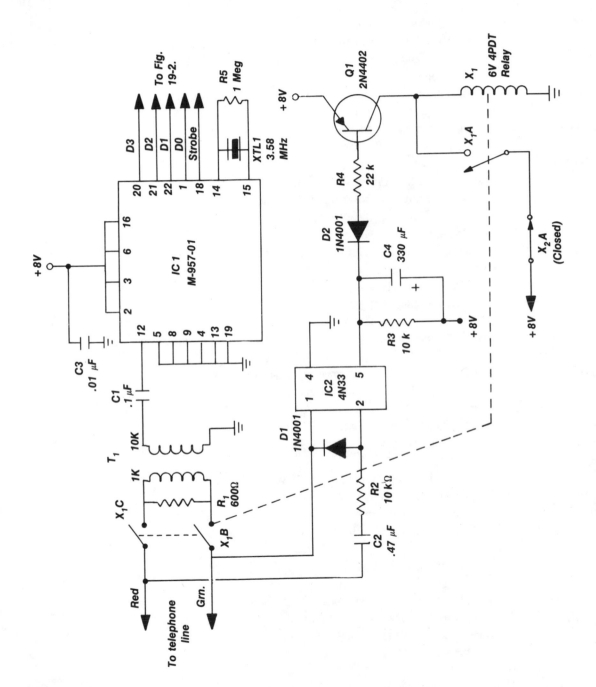

Fig. 19-1. Schematic of the Appliance Controller project.

IC4. At this pin a ground is placed, while all the other outputs of IC4 will remain high (positive voltage).

With a high resistance relay connected to pin 6 the difference in potential between one side of the relay (ground) and the other side (positive voltage), the relay pulls in (see relays X4 and X5 in Fig. 19-2). If the contacts of relay X4 were connected to an ac voltage source, these contacts can control the on and off function of an appliance. The ground on pin 6 will last only as long as IC1 receives the telephone tones for that number, as means of keeping the X4 and the other controller relays on in the absence of a grounding signal. The latching relay approach can also be used here (the latching relay technique was used with relay X1).

When relay X4 closes, contact X4A closes. This closing delivers a grounding voltage through the normally-closed relay contacts of X3. When IC1 does not receive the tones of a number 5 button, the ground at pin 6 is removed but the contacts of X4 have closed providing another path for ground. This second grounding path will keep relay X4 pulled in. IC4 provides a grounding output for 12 telephone push-buttons but only 10 are used to control any appliances. The other two are used for other purposes.

I mentioned before that the caller can control the on and off function of connected appliances. To turn off an appliance, all you must do is press the "*" button on the tone-dial. IC1 decodes this button and IC4 places a ground on pin 13. This ground is applied to a reed-type relay (X3). The contacts of the X3 relay are wired in such a way that through its normally closed contacts (X3A) provides the holding (ground) voltage for all relays that have been turned on by the controller. By pressing the "*" tone-dial button relay X3 pulls in, disconnecting the common ground from all controller relays. With this ground missing, these relays will drop out disconnecting their associated appliances from the 117 volts ac power source.

When all desired appliances have been turned on or off, you must disconnect the telephone controller from the line. To do this, press the "#" button on the tone-dial of the telephone that you are calling from. This will place a temporary ground on pin 14 of IC4. This ground will pull in relay X2. Remember the normally closed contacts of this relay?

Contacts X2A are providing a latching ground for the ring detector relay X1. If the latching ground were disconnected from X1, this relay will drop out, opening the other contacts (X1A and X1B). This would be like placing the handset of a telephone back on its cradle when a call is complete. The transformer T1, and R1 would be disconnected from the telephone line. The appliance controller would then sit patiently for your next call.

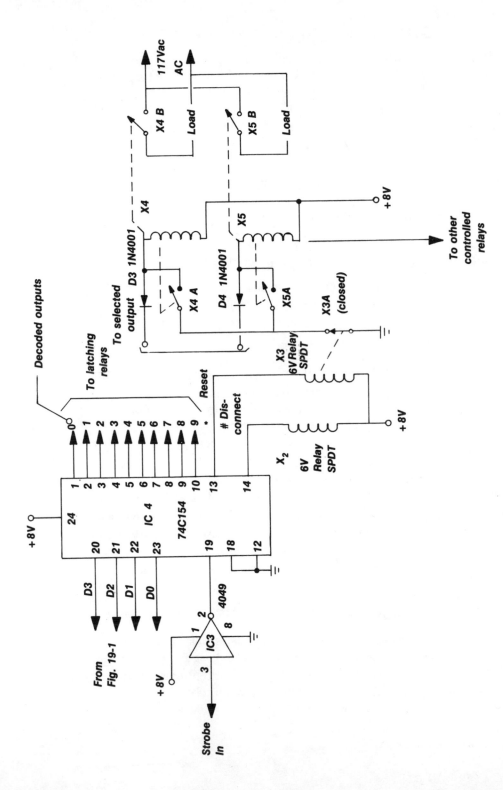

Fig. 19-2. Schematic of the Appliance Controller.

## SELECTING APPLIANCE RELAY NUMBERS

The appliance controller can control up to 10 pieces of equipment. Each appliance relay is turned on only when the selected IC4 output is placed at ground.

If a portable TV is plugged into the controller using relay X4. If you wish to control the TV everytime a tone-dial telephone button number 5 is pressed, solder one side of X4 (see Fig. 19-2 selected output) to IC4 pin 6. If you wish to control a light when you press the number 7, solder relay X5 (light control relay) to pin 8 of IC4.

When IC4 decodes that button number 7 has been pressed, relay X5 will pull in turning on a light. If IC4 decodes that button 5 was pressed, relay X4 will pull in turning on the TV, and so on.

## RELAY COIL RESISTANCE CONSIDERATIONS

IC4 can only handle the grounding of a small amount of current. If a relay were used for X5 that has a coil resistance of only 100 ohms, IC4 cannot pass enough current so that the relay can pull in. To keep cost and wiring to a minimum, all controller relays (X4, X5 . . ., etc.) must have a coil resistance of 1000 ohms or more.

If low-resistance relays are available, you can wire them as in Fig. 19-3. Q1 and Q2 are pnp-type transistors that require a ground to be placed on their base leads so that they conduct. The control relays (X4 and X5) are now connected to the emitter leg of each transistor. When a ground is placed on the base of transistor Q1 by the decoded output of IC4, Q1 conducts, placing a ground on relay X4. This difference of potential will close the relay. The latching and disconnect scheme is the same as discussed earlier.

Using a transistor as a driver will allow you to use relays with a low resistance coil winding. Just remember that the transistor must also be able to handle the extra current needed by the relay. The 2N5139 and 2N4402 transistors can be used on relays with a resistance of about 500 ohms. For lower resistance coils, power transistors must be used in place of the indicated types.

## SCHEMATIC ADDITIONS

To keep the schematic of the appliance controller simple, I neglected to add diodes to all relay coils associated with the circuit. Adding a diode to a relay coil prevents voltage spikes from destroying the ICs. These spikes are produced everytime a magnetic field (like the one generated when relays are energized and de-energized) is created and destroyed. By soldering a diode in the circuit,

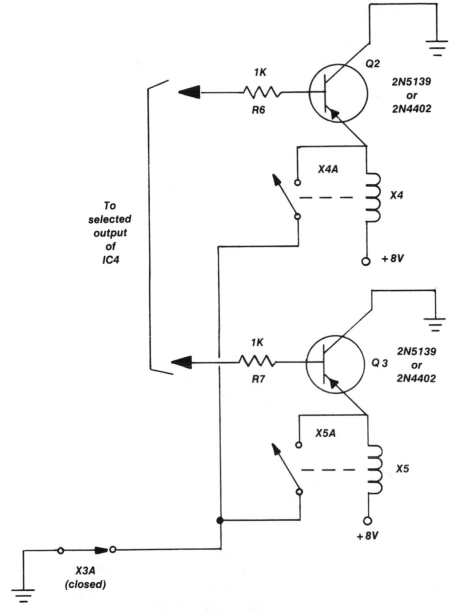

Fig. 19-3. To control relays with low resistance coils, use transistors as drivers.

these spikes are reduced to a point where no harm will come to any other component in the circuit.

Install the diodes by soldering the banded end (of the diodes) to the side of the relay that will receive a positive voltage. The other side of the diode is

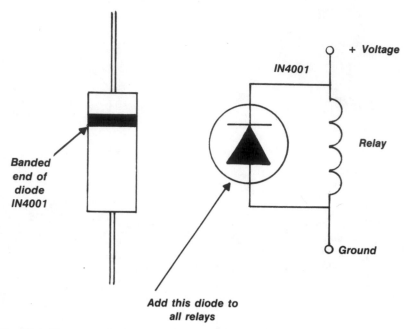

Fig. 19-4. To protect IC and other components from voltage spikes, solder diodes across all relay coils.

to be soldered to the ground terminal of the same relay. Figure 19-4 illustrates the proper way diodes are to be connected. Just remember that diodes have polarities and *must* be inserted correctly.

## ASSEMBLY

The assembly of the appliance controller is relatively simple. Don't let the fact that it took four schematics to present the circuit, worry you. Just use common sense in the purchasing and insertion of the components. Observe proper polarities when inserting the ICs, diodes, transistors, and capacitors.

A perforated board will make the assembly easy, and it should be considered. I am sorry, I did not include artwork for a printed circuit board but I had my reasons. This would only complicate matters by forcing the use of specialized relays and other parts. The PC board would prevent the builder from using spare components that may already be on hand. The artwork of a PC board would be defeating the purpose of this book. To provide high quality electronic projects with little or no cost to the builder.

**NOW A WORD OF CAUTION:** I mentioned it before and I will mention it again, this project makes use of 117 volt ac. This can be deadly. If you are unfamiliar with this type of wiring do not use this

*Parts List*

| | |
|---|---|
| R1 | 600 Ohm Resistor |
| R2 R3 | 10K Ohm Resistor |
| R4 | 22K Ohm Resistor |
| R5 | 1 MegOhm Resistor |
| R6 R7 | 1K Ohm Resistor |
| C1 | .1$\mu$f Capacitor |
| C2 | .47$\mu$f Capacitor |
| C3 | .01$\mu$f Capacitor |
| C4 | 330$\mu$f 16V Capacitor |
| D1 D2 ect. | 1N4001 Diode |
| Q1 Q2 Q3 | 2N4402 or 2N5139 Transistor |
| T1 | Audio Transformer 1K Primary 10K Secondary Res. |
| XTL1 | 3.58 MHz Crystal |
| IC1 | M-957-01 Tone Dial Decoder |
| IC2 | 4N33 Opto-Isolator IC |
| IC3 | 4049 Buffer IC |
| IC4 | 74C154 Decoder IC |
| X1 | 6 volt Relay 4PDT |
| X2 X3 | 6 volt Relay SPDT |
| X4 X5 etc. | 6 volt Relay DPDT - Note that relay contacts must be able to handle the Wattage of the appliance being controlled |

Misc. Parts

| | |
|---|---|
| 1 | Telephone Line Cord |
| 1 | 8 Volt dc Power Supply 12 volt power supply presented in Ch 18 will operate with this circuit. Just change the voltage regulator IC from 7812 to 7808 |

IC1 can be purchased from:
Del. Phone Industries
4487 Plumosa St.
Spring Hill, FL 34606
(after Feb. 1, 1989)

Fig. 19-5. Parts list for the Appliance Controller.

project to control ac appliances. Instead, construct the appliance controller to control low-power dc circuits.

## RELAY CURRENT RATINGS

Another consideration in regards to the control relays (X4 and X5, etc.), determine the amount of current that the appliance you wish to control is drawing when in use. If the appliance draws 100 watts from the line, and you control it through a relay contact that is rated for only 1 or 2 watts, a severe fire hazard will result.

Select a power relay for high current switching. The current capability of a relay can be seen by looking at the width and thickness of its contacts. Contacts that are very wide and thick can handle large currents without any worry to the builder.

## ADD-ON CIRCUITS

With a basic appliance controller assembled, you might want to consider adding a few enhancements of your own. One can be a tone generator that sounds for a second or two to indicate that a relay has been activated. Consider using a different tone for each control relay. Another add-on could be a timer using a 555 integrated circuit. This timer can reset your controller if a telephone call was made by someone other than you. The 555IC will disconnect the system, otherwise the controller would wait for a "#" button to be pressed to disconnect the line.

The 555 IC can be wired in a timing configuration. The timer will start to count down as soon as relay X1 is energized. If by a pre-determined count, no control signals have been received by IC4, the 555 IC (with the help of another normally closed relay wired in series with the X2A contacts) will disconnect a ground signal from relay X1, releasing the controller from the telephone line. These ideas are just that. Ideas. A circuit like this just cries out for modifications that will make it operate in other astonishing ways. The end product is really up to your imagination.

# 20 Animated Telephone Ringer

IN THE EARLY DAYS OF TELEPHONE EQUIPMENT THE ONLY COLOR AVAILABLE WAS black. The only model you could have was a desk-type. And in the beginning, all telephones were rented from the telephone company for a monthly charge. It is said "The more things change. The more things stay the same." In the world of telephones, this saying cannot be any farther from the truth. When the U.S. courts broke up the large AT&T corporation it opened the door wide for Japanese imports. Today, anything goes in the way of telephones. There are telephones built inside plastic footballs. They are in the shape of Star Wars and Disney cartoon figures. These telephones are said to be personalized or decore-type equipment.

You can probably find a telephone to fit the needs or taste of almost any person. With this in mind, I will introduce, in this book, the worlds only Animated Telephone Ringer. And if you have children, you have practically all the needed parts.

The word animation means—to make alive. With this project, you can make your telephone come alive by interfacing a motorized child's toy. You've seen them in toy stores. Small toy bears that bang on a drum when you switch them on. Or other woodland creatures that move with the help of two 1.5 volt batteries.

Figure 20-1 is a photo of the animated telephone ringer that I have. Its a small dog that barks and wags its tail everytime the phone rings. This is a sure-

Fig. 20-1. A photo of my Animated Telephone Ringer. **(Courtesy of Del-Phone Industries Inc.)**

fire way to break the ice in a dull company meeting. Especially when a call comes through and a toy dog that is sitting quietly on the table, starts to bark. It's a great conversation piece.

## HOW IT WORKS

Figure 20-2 shows the interface needed to connect the animated figure to a telephone line. Closer inspection reveals a ring detector circuit using the now famous 4N33 opto-isolator IC. With a 120 volt ac, 30 Hz ring signal applied to the red and green telephone wires, the internal LED of IC1 starts to flash. This flashing allows the Darlington transistor pair (across pins 4 and 5) to conduct, charging capacitor C2 (2.2 $\mu$f 16 volts). It is this capacitor that smoothes or filters the pulsating dc voltage that is developed across pins 4 and 5 of the IC.

This now smooth dc signal is delivered to the base leg of transistor Q1. Q1 is a pnp transistor. This transistor needs a ground (or negative voltage) on its base so that it can conduct. It is this conducting transistor that allows the X1 relay to pull in.

The on/off cycle of a ring signal is 1 second on (ringing) 3 to 4 seconds off (no ring). The way the interface is presented in Fig. 20-2, the relay will drop

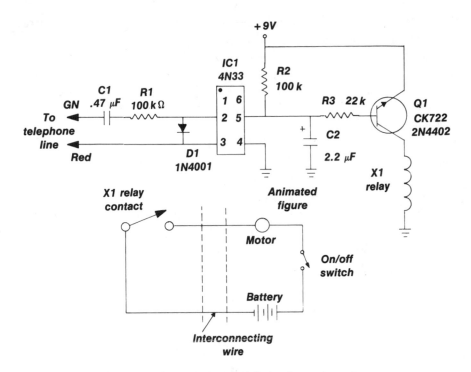

Fig. 20-2. Schematic of the Animated Telephone interface.

**Parts List**

| | |
|---|---|
| R1 R2 | 100K Ohm Resistors |
| R3 | 22K Ohm Resistor |
| C1 | .47μf 250V Capacitor |
| C2 | 2.2μf or 330μf Capacitor<br>See Text |
| Q1 | CK722 or 2N4402 Transistor |
| X1 | SPST 6 volt Relay |
| IC1 | 4N33 Opto-Isolator IC |
| D1 | 1N4001 Diode |
| Misc. Parts | |
| 1 | Telephone Line Cord |
| 1 | Motorized Animated Toy Figure<br>(See Text) |

Fig. 20-3. Parts list for the Animated Telephone Ringer.

out on every 3 to 4 second off cycle. If you wish the animated figure to continue operating even during this off cycle, replace capacitor C2 with a 330 μf 16 volts. This larger value capacitor will discharge a stored voltage into the base of transistor Q1 every time off cycle is encountered. This slow discharging will turn on the animated figure, and it will continue to operate until you answer the telephone, at that time the ring signal will stop but the stored voltage across

C2 must be discharged before transistor Q1 stops conducting. This extra discharging time is about 4 to 4½ additional seconds. This means that the animated figure will continue to operate for an additional 4 to 4½ seconds after you pick up the telephone receiver.

The interface electronics can be wired to the figure by using a pair of very fine wires. I found that the wire used on an inexpensive transistor radio earphone will work just fine.

You must choose the type of animated figure to be interfaced with the detector. All that is needed for the figure to swing into action when the telephone rings is a pair of wires (earphone wires) that have been connected in series with the figure's motor, battery, and on/off switch. This connection can be made directly inside the battery compartment by cutting one wire from the battery

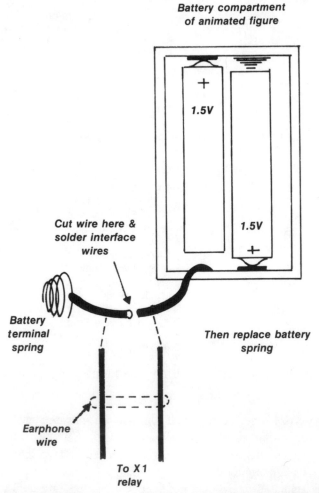

Fig. 20-4. How to connect the animated toy figure to the interface.

spring, and soldering this wire to one of the interface cable wires. The second wire from the interface is then soldered to the battery spring. This wiring arrangement can be seen in Fig. 20-4.

When the interface cable is connected to the figure solder the other end to the normally open contacts of relay X1. When the relay closes (indicating an incoming telephone call) the series circuit of the animated figure is complete, activating the figures' internal motor. The figure will then move, bark, wag its tail, or just about anything it's supposed to do.

## ASSEMBLY

There is no critical area in the assembly. To simplify the assembly even further, Fig. 20-5 contains prepared PC board artwork. Figure 20-6 shows parts placement for the board if it is made. Care must be taken in the insertion of polarity sensitive components such as integrated circuits, electrolitic capacitors, and transistors. If you pay attention to proper assembly techniques there should be no problem in building an operational project. With your Animated Telephone Ringer assembled, you can test its operation by shorting pins 4 and 5 of IC1 together. This will by-pass the high internal resistance associated with these terminals. When shorted, the X1 relay should pull in and turn on the motorized

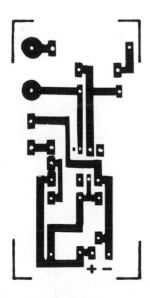

*Animated telephone
ringer*

Fig. 20-5. PC board artwork for the interface used with the Animated Telephone Ringer.

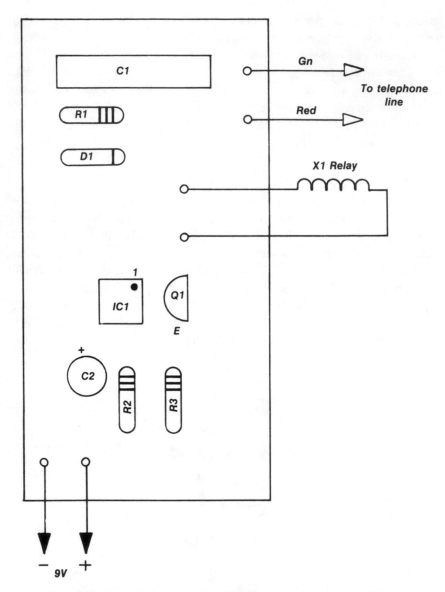

Fig. 20-6. Component layout for the Animated Telephone Ringer.

figure. Then remove the short. If C2 is the intended 2.2 $\mu$f, the figure should stop as soon as the short is removed. If C2 was replaced with the 330 $\mu$f capacitor, the figure should continue to operate 4 to 4½ seconds after the short is removed.

## A FINAL NOTE

This circuit concludes my presentation of the 15 Telephone Enhancement Projects for the Hobbyist. I hope you built many of the enclosed projects and were happy with the results. Because I was happy to bring to you a small portion of my lifes work.

# Index

# Index

**162**

# Other Bestsellers From TAB

☐ **MASTERING HOUSEHOLD ELECTRICAL WIRING—2nd Edition—James L. Kittle**

Update dangerously old wiring in your house. Add an outdoor dusk-to-dawn light. Repair a malfunctioning thermostat and add an automatic setback. You can do all this and more—easily and safely—for much less than the cost of having a professional do it for you! You can remodel, expand, and modernize existing wiring correctly and safely with this practical guide to household wiring. From testing to troubleshooting, you can do it all yourself. Add dimmer switches and new outlets . . . ground your TV or washer . . . make simple appliances repair . . . set up outside wiring . . . put in new fixtures and more! 304 pp., 273 illus.
**Paper $18.95**                               **Hard $24.95**
**Book No. 2987**

☐ **THE ILLUSTRATED HOME ELECTRONICS FIX-IT BOOK—2nd Edition—Homer L. Davidson**

This revised edition of the bestselling home electronics fix-it handbook will save you time and aggravation AND money! It is the only repair manual you will ever need to fix most household electronic equipment. Packed with how-to illustrations that any novice can follow, you'll soon be able to fix that broken television and portable stereo/cassette player and "Boom Box" and intercom and . . . the list goes on! 480 pp., 377 illus.
**Paper $19.95**                               **Hard $25.95**
**Book No. 2883**

☐ **TROUBLESHOOTING AND REPAIRING AUDIO EQUIPMENT—Homer L. Davidson**

When your telephone answering machine quits . . . when your cassette player grinds to a stop . . . when your TV remote loses control . . . or when your compact disc player goes berserk . . . You don't need a degree in electronics or even any experience. Everything you need to troubleshoot and repair most common problems in almost any consumer audio equipment is here in a servicing guide that's guaranteed to save you time and money! 336 pp., 354 illus.
**Paper $18.95**                               **Hard $24.95**
**Book No. 2867**

☐ **HOME PLUMBING MADE EASY: AN ILLUSTRATED MANUAL—James L. Kittle**

Here, in one heavily illustrated, easy-to-follow volume, is all the how-to-do-it information needed to perform almost any home plumbing job, including both water and waste disposal systems. And what makes this guide superior to so many other plumbing books is the fact that there's plenty of hands-on instruction, meaningful advice, practical safety tips, and emphasis on getting the job done as easily and professionally as possible! 272 pp., 250 illus.
**Paper $12.95**                               **Hard $14.95**
**Book No. 2797**

☐ **TROUBLESHOOTING AND REPAIRING SMALL HOME APPLIANCES—Bob Wood**

Author Bob Wood pairs step-by-step pictures with detailed instructions on how to fix 43 of the most common electric appliances found in the home. Following the illustrations and directions provided, you'll be able to quickly disassemble practically any electrical device to get to the trouble source. Among those included are: drill, garbage disposal, can opener, grass trimmer, vacuum cleaner, blender, and much more! Telltale symptoms, troubleshooting techniques, maintenance measures—even operating tests and instruments—are included for each fix-it project featured. 256 pp., 473 Illus.
**Paper $18.95**                               **Hard $23.95**
**Book No. 2912**

☐ **AIR CONDITIONING AND REFRIGERATION REPAIR—2nd Edition—Roger A. Fischer and Ken Chernoff**

The second edition of *Air Conditioning and Refrigeration Repair* covers the basics of electricity, refrigeration theory, controls, refrigerants, water chemistry, soldering, and charging the units, as well as the latest developments in the field itself. The authors explain in a nontechnical manner the features, care, and repair of all components of a refrigeration unit. 384 pp., 229 illus.
**Paper $19.95**                               **Hard $26.95**
**Book No. 2881**

☐ **THE CAMCORDER HANDBOOK—Gerald V. Quinn**

Takes you step-by-step through the entire production process using a camcorder. Going well beyond what is included in the standard owner's manual, Quinn provides guidance in lighting, sound, camera movement, and more—for shooting indoors or out! Whether you're taping a school play, sporting event, wedding, vacation, children, recording a family history, or making a tape for business purposes, you'll have the basic understanding of video concepts and techniques that you need. 240 pp., 139 illus.
**Paper $14.95**                               **Hard $18.95**
**Book No. 2801**

☐ **COMPACT DISC PLAYER MAINTENANCE AND REPAIR—Gordon McComb and John Cook**

Packed with quick and reliable answers to the problems of maintaining and repairing CD players, this illustrated, do-it-yourself guide takes the apprehension out of first-time repairs. The authors take away the mystery that surrounds these seemingly complicated devices and give you the confidence you need to repair minor malfunctions (the cause of more than 50% of CD player problems). 256 pp., 188 illus.
**Paper $14.95**                               **Hard $18.95**
**Book No. 2790**

# Other Bestsellers From TAB

☐  **MAJOR HOME APPLIANCES: A Common Sense Repair Manual—Darell L. Rains**

Prolong the life and efficiency of your major appliances . . . save hundreds of dollars in appliance servicing and repair costs . . . eliminate the frustration of having to wait days, even weeks, until you can get a serviceman in to repair it! With the help and advice of service professional Darell L. Rains, even the most inexperienced home handyman can easily keep any washer, dryer, refrigerator, icemaker, or dishwasher working at top efficiency year after year! 160 pp., 387 illus.

**Paper   $16.95**                                    **Hard   $21.95**
**Book No. 2747**

☐  **THE COMPLETE HANDBOOK OF VIDEOCASSETTE RECORDERS—3rd Edition—Harry Kybett and Delton T. Horn**

This handbook is a must-have selection for anyone who owns, uses, or is thinking of purchasing a VCR! Now, completely revised and updated, it explores the latest VHS, Beta, and 8MM systems including the Sony 5000 series, SMPTE type C 1″ format, and Beta CAM and M format—as well as older machines. 256 pp., 254 illus.

**Paper   $16.95**                                    **Hard   $21.95**
**Book No. 2731**

**Send $1 for the new TAB Catalog describing over 1300 titles currently in print and receive a coupon worth $1 off on your next purchase from TAB.**

(In PA, NY, and ME add applicable sales tax. Orders subject to credit approval. Orders outside U.S. must be prepaid with international money orders in U.S. dollars.)

*Prices subject to change without notice.

To purchase these or any other books from TAB, visit your local bookstore, return this coupon, or call toll-free 1-800-233-1128 (In PA and AK call 1-717-794-2191).

| Product No. | Hard or Paper | Title | Quantity | Price |
|---|---|---|---|---|
|  |  |  |  |  |
|  |  |  |  |  |
|  |  |  |  |  |

☐ Check or money order enclosed made payable to TAB BOOKS Inc

Charge my   ☐ VISA    ☐ MasterCard    ☐ American Express

Acct. No. _____ Exp. _____

Signature _____

Please Print
Name _____

Company _____

Address _____

City _____

State _____ Zip _____

| | |
|---|---|
| Subtotal | |
| Postage/Handling ($5.00 outside U.S.A. and Canada) | $2.50 |
| In PA, NY, and ME add applicable sales tax | |
| TOTAL | |

Mail coupon to:

**TAB BOOKS Inc.**
Blue Ridge Summit
PA 17294-0840                          BC